Embedded Systems

Volume 20

Series Editors

For further volumes:
http://www.springer.com/series/8563

Alberto Sangiovanni-Vincentelli
Haibo Zeng · Marco Di Natale
Peter Marwedel
Editors

Embedded Systems Development

From Functional Models to Implementations

Editors
Alberto Sangiovanni-Vincentelli
Department of Electrical Engineering and Computer Science
University of California
Berkeley, CA
USA

Haibo Zeng
Department of ECE
McGill University
Montreal
Canada

Marco Di Natale
TeCIP Institute
Scuola Superiore Sant'Anna
Pisa
Italy

Peter Marwedel
Embedded Systems Group
TU Dortmund University
Dortmund
Germany

ISSN 2193-0155 ISSN 2193-0163 (electronic)
ISBN 978-1-4614-3878-6 ISBN 978-1-4614-3879-3 (eBook)
DOI 10.1007/978-1-4614-3879-3
Springer New York Heidelberg Dordrecht London

Library of Congress Control Number: 2013939569

Printed on acid-free paper

Springer is part of Springer Science+Business Media (www.springer.com)

Preface

This book is an edited collection of contributions in selected topics related to embedded systems modeling, analysis, and synthesis. Most contributions are extended versions of papers that were originally presented at several workshops organized in the context of the Embedded Systems Week and Real-Time Systems Symposium in the last months of 2011. The workshops targeted topics and challenges related to the use of models for the design, analysis, and synthesis of Embedded Systems. These workshops were the *WSS*, *Workshop on Software Synthesis*, the *TiMoBD*, *Time Analysis and Model-Based Design*, and the *SOMRES*, *Workshop on Synthesis and Optimization Methods for Real-Time Embedded Systems*. Problems and solutions were discussed for different stages in the development process and applied to the system-level view, as well as to the design, analysis, and synthesis of components and subsystems and the behaviors therein.

As workshop organizers and editors of this book, we believe that recent years have brought renewed interest for the study and development of embedded and cyber-physical systems by researchers and developers. The opportunity for the development of new languages, methods, and tools comes from the emergence of feature-rich, complex, distributed systems, and the need to tame their complexity in new ways, is leading to the adoption of model-based development, new analysis methods and design synthesis techniques, and true component-based development, in which functional and platform assemblies are composable, analyzable, and, possibly in the future, demonstrably correct-by-construction.

This book collects contributions on different topics, including system and software models, innovative architectures (including OS and resource managers), formal methods, model checking and analysis techniques, software synthesis, system optimization and real-time networks, with the ambitious objective of providing useful insights and innovative ideas on how to solve very complex problems throughout the entire (model-based) development cycle. Contrary to other books on the subject, we attempt at reconciling the two communities of Model-Based Design and Model-Driven Engineering, which often operate in independent ways, with only a few fortunate exceptions.

Regardless of the workshop organization, the selected papers have been reorganized according to their topics and divided into parts that better fit the stages in the development process rather than an abstract classification based, for example, on languages, algorithmic solutions, or analysis and synthesis methods. The intended audience includes of course the general community of embedded systems researchers, but we believe several topics should be also of interest for developers, tool vendors, and development process experts. Several contributions are provided by industry developers and researchers, referring to upcoming commercial products, methods, and tools. The applicability of most other results is demonstrated by use cases and/or project experiences.

We would like to acknowledge all authors for their hard work, the reviewers and workshop audiences for their constructive feedback and interesting discussion, which eventually paved the way for improved and new content, and the assistant editors at Springer. We hope that this book will serve as an interesting source of inspiration for new researches and applications, and help many readers to enter the domain of model-based design of embedded systems.

Berkeley-Rome, March 2013 Alberto Sangiovanni-Vincentelli
Montreal Haibo Zeng
Pisa Marco Di Natale
Dortmund Peter Marwedel

Contents

Chapter 1
Introduction: Modeling, Analysis and Synthesis of Embedded Software and Systems

Alberto Sangiovanni-Vincentelli, Haibo Zeng, Marco Di Natale and Peter Marwedel

Abstract Embedded systems are increasingly complex, function-rich and required to perform tasks that are mission- or safety-critical. The use of models to specify the functional contents of the system and its execution platform is today the most promising solution to reduce the productivity gap and improve the quality, correctness and modularity of software subsystems and systems. Models allow to advance the analysis, validation, and verification of properties in the design flow, and enable the exploration and synthesis of cost-effective and provably correct solutions. While there is (relative) consensus on the use of models, competing (and not necessarily compatible) approaches are explored in the academic and industrial domain, each with its distinctive features, strengths, and weaknesses. Modeling languages (and the accompanying methodologies) are today roughly divided as belonging to the Model-Based Design (MBD) or Model-Driven Engineering (MDE) approach. Component-based development is a desirable paradigm that applies to both modeling styles. Research work tries to define (and possibly widen) the range of model properties that can be analyzed and demonstrated as correct, providing methods and tools to this purpose. Time properties are an important subset, since they apply to the majority

A. Sangiovanni-Vincentelli (✉)
Department of Electrical Engineering and Computer Science (EECS),
University of California, Berkeley, 253 Cory Hall MC# 1770,
Berkeley, CA 94720-1770, USA
e-mail: alberto@eecs.berkeley.edu

H. Zeng
Department of ECE, McGill University, Montreal H3A 2A7, Canada
e-mail: haibo.zeng@mcgill.ca

M. Di Natale
TeCIP Institute, Scuola Superiore Sant'Anna, via Moruzzi 1, Pisa 56124, Italy
e-mail: marco.dinatale@sssup.it

P. Marwedel
Embedded Systems Group, TU Dortmund University, Otto-Hahn-Str. 16,
Dortmund 44221, Germany
e-mail: peter.marwedel@tu-dortmund.de

A. Sangiovanni-Vincentelli et al. (eds.), *Embedded Systems Development*,
Embedded Systems 20, DOI: 10.1007/978-1-4614-3879-3_1,

of the complex and distributed systems in the automotive, avionics, and controls domains. A synthesis path, with the methods and tools to generate a (provably correct) software or hardware implementation of a model is a necessary complement to the use of an analyzable modeling language, not only to improve efficiency, but to avoid the introduction of unwanted errors when the model is refined into its implementation.

Embedded systems are pervasive in today's society, and everybody can witness the increasing number of intelligent functions in everyday devices. What is probably less evident is the number of problems faced by the industry in coping with the increasing complexity of embedded functionality. The need for more contents and new intelligent functions is openly in a mismatch with the request for a reduced time-to-market, more robustness, and future extensibility.

Traditional programming techniques, including object-oriented languages, are not able to reduce the productivity gap, and embedded system development processes demand new methods and techniques that can improve the quality, correctness, and modularity of systems and subsystems by advancing the analysis and verification of properties as early as possible in the design flow.

The use of models can help the analysis of the system properties and verification by simulation, the documentation of the design decisions, and possibly the automatic generation of the software implementation. Each of the previous topics is the subject of a number of relevant research domains, but all of them are also part of the industrial practice, at least to some degree, backed by several commercial products and standards.

The Model-Based Design approach (MBD) prescribes the use of models based on a mathematical formalism and executable semantics to represent the controller system (to be realized in SW or HW) and the controlled device or environment (often referred to as *Plant*). Examples of available commercial tools for model-based development are Simulink [1], SCADE [2], NI LabVIEW [3], and Modelica [4]. Academic projects that fit this definition are Ptolemy [5] and Metro II [6].

These tools are feature-rich and allow the modeling of continuous or discrete time, or hybrid systems in which functionality is typically represented using a dataflow or an extended finite-state machine formalism (or a combination of them). MBD languages enable the realization of mathematical specifications that can be used for system simulation, testing and behavioral code or firmware generation. Very often these languages and tools tend to be domain-specific; subtle semantic differences may characterize each of them. The way in which time, time events, and time specifications are handled by these languages is a main variation point. In many cases, especially when an implementation is automatically generated from a model, time events belong to a discrete lattice and all time properties are expressed in terms of multiples of the system base rate. For discrete-time models, several analysis and verification techniques that are typical of hardware design can be reused. When continuous-time modeling is required and controller events can happen at any point in time, timed automata [7] models can provide an adequate formal backing and verification capability.

The strength of MBD tools is the analysis, synthesis and verification capability. However, when the software/hardware implementation of models needs to be considered at simulation time to account for computation and communication delays, or at code generation time, most commercial MBD tools lack expressive power to represent complex system architectural aspects and execution platforms. Moreover, their management of types, components and extension mechanisms are often limited in scope.

Originated from the software engineering research community, Model-Driven Engineering (MDE) is a methodology which focuses on creating and exploiting domain specific models with abstract representations of the knowledge and activities in a particular application domain. The Model-Driven Architecture (MDA) [8] initiative was started by the Object Management Group (OMG) as the driver for the use of models in a process where the functional description is separated from the platform model and later bound together in the development stages. The binding of the two models and the generation of a platform-specific model (PSM) is to be achieved through suitable model-to-model transformations. The languages prescribed by the OMG for the MDA approach derive from experiences in the (object-oriented) software engineering. These (graphical) languages are suited at representing architectural aspects and are designed for extensibility. In addition, OMG provided standard means for the definition of metamodels and standard languages for model-to-model transformations and model-to-text (model-to-code) generation. However, the behavioral semantics of the languages recommended by OMG for the MDA (including UML [9], and SysML [10], purposely defined for modeling systems and embedded systems) had not been completely and formally specified until very recently, with the new OMG Action Semantics language ALF [11] (the semantics is specified as fUML [12]). Thus, model execution and simulation are today tool-specific and often characterized by a limited scope.

The extensibility features of these languages have been exploited for the definition of a domain-specific extension for Real-time and Embedded systems, called MARTE [13]. MARTE includes a set of specialized definitions for time, timed events, time attributes such as worst-case execution times, and time constraints. It is today the natural candidate for the definition of large-scale embedded systems with real-time constraints under the MDE approach. Several groups proposed the use of worst-case timing analysis techniques, working at the level of software tasks and messages to analyze the correctness of systems modeled using UML and SysML with MARTE.

Model-Driven Engineering (MDE) and Model-Based Development (MBD) are possible choices to form the backbone of the design flow. Albeit MDE and MBD share the same principle (using models as primary artifacts to drive the design and development process), they differ substantially in their details. Each on its own is incapable of addressing all the challenges of modern system design. However, their strengths and weaknesses can be seen as complementary, and they can be complemented by domain-specific languages or other formal models. Among those, Architecture Description Languages or ADL for the architecture-level views and Timed-Automata for the description of the behavior and the formal verification of timing properties are proposed in one of our featured contributions.

Book Outline

The book chapters are organized around four main subjects. The first part focuses on model-based design languages and tools with emphasis on the analysis (by simulation and early prototyping), and especially synthesis capabilities offered by these languages. The chapters in this part discuss dataflow models, languages, and tools, and their application to streaming and signal-processing applications.

The second part discusses the languages belonging to the Model-Driven Engineering approach and domain-specific architecture-description languages. These languages are used to describe, analyze, and demonstrate time properties of systems, perform synthesis of software and hardware (at least under some restrictions), and integrate the architecture-level description with formal descriptions of the behavior (using timed automata) to demonstrate properties of interest by model-checking.

The third part focuses on components, composability, and compositionality. Composability means that the properties of a component are preserved when the component is connected to or operates in conjunction with other components. Compositionality means that the properties of an assembly of components (or composite) can be analyzed starting from the properties of the atomic components. The contributions to this part are largely independent from the specific language that is used to define the component structure and behavior, but rather focus on the properties that such languages should have, and the required assumptions and assertions on each component for each property of interest. Time properties play an important role and are the subject of two of the three contributions.

Finally, the subject of the fourth part is timing analysis in the general sense. In this part, the definition and use of domain-specific languages for the description of timing-properties in large-scale embedded systems are discussed. Also, analysis and synthesis techniques for the optimal assignment of architecture-level parameters and attributes (task and message priorities) are presented. Timing analysis (as discussed in the very last chapter of this part) should not be a tool simply for verifying the correctness of a design solution to a real-time specification, but is highly valuable for the selection of the best platform support and architecture-level solution to the specification problem in the architecture evaluation and selection stage.

1.1 Recommended Reading

The purpose of this section is to provide the background and references that are necessary for reading the book chapters and catching up on some concepts, terms, languages, and methods that may be assumed as part of the readers' background. In addition, we take this opportunity to provide an overview of other results and research projects that we consider important to have a comprehensive view of the state of the art in embedded system development and the use of models therein.

The separation of the two main concerns of functional and architectural specification and the mapping of functions to architecture elements are among the founding principles of many design methodologies such as Y-chart [14], platform-based design

(PBD) [15] , Model-Driven Engineering MDE [16], and tools like the Ptolemy [5] and Metropolis [17] frameworks. In the MDE staged development, a Platform Independent Model or PIM is transformed into a Platform Specific Model (PSM) by means of a Platform Definition Model (PDM). The match of a functional and execution architecture is advocated by many in the academic community and in the industry as a way of obtaining modularity and separation of concerns between functional specifications and their implementation on a target platform. It is therefore also the foundation of emerging standards and recommendations, such as the UML MARTE [13] profile from the Object Management Group (OMG) [18] , as well as the best practices in industry, such as the V-cycle of software development [19] and the AUTOSAR standard [20] from the automotive industry.

A short review of the most common models of computation (formal languages) proposed by academia and possibly supported by the industry with the objective of formal or simulation-based verification is necessary for understanding opportunities and challenges in the use of models for the development of embedded and cyber-physical systems (please refer to [21, 22] for more detail).

Formal models are mathematical-based languages that specify the semantics of computation and communication (also defined as Model of Computation or MOC [23]). MOCs may be expressed, for example, by means of a language or automaton formalisms. A system-level MOC is used to describe the system as a (possibly hierarchical) collection of design entities (blocks, actors, tasks, and processes) performing units of computations represented as transitions or actions, characterized by a state and communicating by means of events (tokens) carried by signals. Composition and communication rules, concurrency models, and time representation are among the most important characteristics of an MOC.

Once the system specifications are given according to a formal MOC, formal methods can be used to achieve design-time verification of properties and implementation. In general, properties of interest go under the two general categories of Ordered Execution and Timed Execution: Ordered Execution attains to verification of event and state ordering. Properties such as safety, liveness, absence of deadlock, fairness, and reachability belong to this category. Timed Execution attains to event enumeration, such as checking that no more than n events (including timed events) occur between any two events in the system. Timeliness and some notions of fairness are examples of this type. Verification of some desirable system properties may be quite hard or even impossible to achieve by logical reasoning on formal models. Formal models are usually classified according to the decidability of properties (the interested reader may refer to [22] for a survey on the subject). In practice [21], decidability should be carefully evaluated. In some cases, even if it is decidable, the problem cannot be practically solved since the required run-time may be prohibitive; in other instances, even if undecidability applies to the general case, it may happen that the problem at hand admits a solution.

In Finite State Machines (FSM), process behavior is specified by enumerating the (finite) set of system states and the transitions among them. In the Synchronous FSM model, signal propagation is assumed to be instantaneous, and transitions and the evaluation of the next state happen for all the system components at the same time. Synchronous languages [24], such as Esterel [25, 26] and Lustre [27, 28] and, to some degree, the Simulink graphical language are based on this model. In the Asynchronous FSM model, two asynchronous FSMs never execute a transition at the same time except when explicit rendezvous is specified (a pair of transitions of the communicating FSMs occur simultaneously).

In the Statecharts extension of FSMs [29], Harel proposed three mechanisms to reduce the size of a Finite State Machine for modeling practical systems: state hierarchy, simultaneous activity, and non-determinism. In Petri Net (PN) models, the system is represented by a graph of places connected by transitions. Places represent unbounded channels that carry tokens, and the state of the system is represented at any given time by the number of tokens existing in a given subset of places. Transitions represent the elementary reaction of the system.

The FSM and PN models have been originally developed with no reference to time or time constraints, but the capability of expressing and verifying timing requirements is key in many design domains (including embedded systems). Hence, both have been extended in order to allow time-related specifications. Time extensions differ according to the time model that is assumed. Models that represent time with a discrete time base are said to belong to the family of discrete time models, while the others are based on continuous (dense) time. Many models have been proposed in the research literature for time-related extensions. Among those, Time Petri Nets (TPN) [30] and Timed Automata [7] are probably the best known. Timed automata (TA) operate with a finite set of locations (states) and a finite set of real-valued clocks. All clocks proceed at the same rate and measure the amount of time that passed since they were started (reset). Each transition may reset some of the clocks. It also defines a set of restrictions on the value of the symbols as well as on the clock values. A few tools based on the timed automata models have been developed. Among those, we cite Kronos [31] and Uppaal [32]. Kronos is a tool developed to check whether a timed automaton satisfies requirements expressed using the real-time temporal logic TCTL. The Uppaal tool allows modeling, simulation and verification of real-time systems modeled as a collection of non-deterministic processes with finite control structure and real-valued clocks, communicating through channels or shared variables. The tool is free for no profit and academic institutions.

The UML Unified Modeling Language was developed in the context of general purpose computing as the merger of many object-oriented design methodologies aimed at the definition of generic software systems. Its semantics intentionally retains many variation points in order to adapt to different application domains. To be practically applicable to the design of embedded systems, further characterization (a specialized profile in UML terminology) is required. In the revision 2.0 of the language, the system is represented by a (transitional) model where active and passive components, communicating by means of connections through port interfaces, cooperate in the implementation of the system behavior. As a joint initiative between OMG and INCOSE (International Council on Systems Engineering), the SysML language has been defined as an UML extension to provide the additional features that are needed for system-level modeling (UML was developed for modeling software systems only). Clearly, the development of SysML was strongly motivated by the emergence of embedded and cyber-physical systems.

In UML and SysML, the development of a platform model for (possibly large and distributed) embedded systems and the modeling of concurrent systems with resource managers (schedulers) require domain-specific concepts. The OMG MARTE [13] standard is very general, rooted on UML/SysML and supported by several tools. MARTE has been applied to several use cases, most recently to automotive projects [33]. However, because of the complexity and the variety of modeling concepts it has to support, MARTE can still be considered as work in progress, being constantly evaluated [34] and subject to future extensions. Several other domain-specific languages and architecture description languages of course exist, for example, EAST-ADL [35] and the DoD Architectural Framework [36].

Several other authors [37] acknowledge that future trends in model engineering encompass the definition of integrated design flows exploiting complementarities between UML [9] or SysML [10] and Matlab/Simulink [1], although the combination of the two models is affected by the fact that Simulink lacks a publicly accessible meta-model [37].

Work on the integration of UML and synchronous reactive languages [24] has been performed in the context of the Esterel language (supported by the commercial SCADE tool), for which transformation rules and specialized profiles have been proposed to ease integration with UML models [38].

With respect to the general subject of model-to-model transformations and heterogenous models integration, several approaches, methods, tools, and case studies have been proposed. Some proposed methods, such as the GME framework [39], consist of the use of a general meta-model as an intermediate target for the model integration. In Metropolis [17], an abstract semantics is defined to yield a metamodel that has precise properties and can be used as an intermediate target as well but with formal guarantees for analysis and integration [40].

Other projects [5] have developed the concept of studying the conditions for interface compatibility between heterogeneous models. Examples of formalisms developed to study compatibility conditions between different Models of Computation are the Interface Automata [41] and the Tagged Signal Model [42].

If specification of functionality aims at producing a logically correct representation of system behavior, architecture-level design is where physical concurrency and schedulability requirements are expressed. At this level, the units of computation are the processes or threads. Formal models, exhaustive analysis techniques, and model checking are now evolving towards the representation and verification of time and resource constraints together with the functional behavior. However, applicability of these models is strongly limited by state space explosion. In this case, exhaustive analysis and joint verification of functional and non-functional behavior can be sacrificed for the easier goal of analyzing only the worst case timing behavior of coarse-grain design entities representing concurrently executing threads. Fixed Priority Scheduling and Rate Monotonic Analysis (RMA) [43, 44] are by far the most common real-time scheduling and analysis methodologies. RMA provides a very simple procedure for assigning static priorities to a set of independent periodic tasks together with a formula for checking schedulability against deadlines on a single processor. Their extensions to distributed systems depend on the semantics of information passing among tasks and messages. Common models are communication by sampling [45], in which periodic tasks and messages exchange information through (protected) shared variables (the analysis and possible synthesis of optimal architecture configurations is discussed in [46, 47]), the holistic, or jitter propagation model [48], in which tasks are activated by the arrival of messages and messages are queued at the completion of a sender task, and the transactional model with activation offsets [49]. A combination of the previous two models is discussed in [50], and the analysis models are nicely summarized in [51].

The following sections provide an introduction to the topics of the book chapters, with a summary discussion of the involved issues and solutions and additional background information.

1.2 Model-Based Design and Synthesis

Model-based design flows are based on the use of models with a mathematical backing and a strong semantics characterization. These models (dataflows and extended, hierarchical state machines are by far the most common formalisms) are often also characterized by an executable semantics and allow for simulation, formal verification of properties on the model and, most often, an available path to implementation.

However, the timing behavior of the system depends on features of the computation and communication architecture that are modeled later or not modeled at all, resulting in the possibility for an inappropriate selection of the computing platform (over- or under-performing) and an incorrect software implementation of the functional model. To this end, simulation and timing analysis techniques can provide support for the analysis of architecture solutions and system configurations, as well as the synthesis of feasible/correct solutions.

The verification of the model properties will become a significant asset with the increasing need for safety and mission-critical systems, where certification is required with the extensive coverage of decisions/conditions (as imposed by the industrial safety standards DO-178B [52] and ISO26262 [53]), which is going to be impractical, because of combinatorial explosion, if performed at test time.

Automated verification by theorem proving/model checking seems to be the only option, but certified model verification requires demonstrated semantics preservation in the generation of refinements/implementations. Unfortunately, the available paths to implementation are de-facto for single-core platforms only, or at most time-triggered platforms and the generated code is statically scheduled or adheres to a very simple task model.

Typically, not many commercial tools provide adequate means to represent the execution platform and the task/message models. Large-scale embedded/Cyber-Physical Systems (CPS) might require the platform modeling and the modeling of the mapping of functionality to platform where the platform is not necessarily time-triggered (a formalization of the concept of a Loosely Time-Triggered Architecture, or LTTA, can be found in [54]). Platform modeling may allow for virtual prototyping and the early evaluation of the timing properties and the feasibility of the model implementation. In one step further, programmable hardware (FPGAs) can be used for the rapid prototyping of systems to be implemented in HW.

The Chap. 2 (*Modeling, Analysis, and Implementation of Streaming Applications for Hardware Targets*) describes a new framework, developed at National Instruments R&D for the development of streaming applications. The framework supports a high-level model or specification based on the static dataflow model of computation and its extensions for cyclo-static data rates and parameterization, the early prototyping and performance estimation on FPGAs, and the synthesis of a HW or SW implementation. The effectiveness of the approach is demonstrated with a case study of an OFDM transmitter and receiver.

The Chap. 3 (*Dataflow-based, Cross-platform Design Flow for DSP Applications*) presents a design methodology based on the core functional dataflow (CFDF) model of computation, where actors must proceed deterministically to one particular mode of execution whenever they are enabled. It supports efficient design space exploration on a variety of platforms including graphics processing units (GPUs), multicore programmable digital signal processors (PDSPs), and FPGAs. It uses a "minimalist" approach for integrating coarse grain dataflow programming structures into DSP simulation, and the dataflow interchange format to provide cross-platform actor design support.

1.3 Model-Driven Design, Integration and Verification of Heterogeneous Models

Model-Driven Engineering encompasses both a set of tools and a methodological approach to the development of software. The software engineer is trying to automate the processes by building and using abstractions, and the produced software will be of better quality than by using a general purpose programming languages. The reasoning behind this claim is that abstractions of concepts and of processes manipulating those concepts are easier to understand, verify, and simulate than computer programs, as those abstractions are close to the domain being addressed by the engineers. Of course, this holds only when the used abstractions are suitable to describe the addressed domain.

To apply such a development flow to the domain of real-time embedded systems, there is a need for better integrations of analysis and verification technologies for real-time embedded systems with methods and tools in model-based and model-driven development flows. In particular, domain-specific concepts that support the development of a platform model for embedded systems and the modeling of concurrent systems with resource managers are required. For example, the UML and SysML modeling languages have been extended to support model-driven development of real-time and embedded application, which is documented in the OMG standard MARTE.

A typical issue with the model-based design and model-driven engineering approaches, or in particular, the integration of real-time analysis and synthesis methods and tools with them, is the gap between different behavioral models used during different development stages by different experts. For example, model transformation techniques are needed to transform UML and SysML (or other architecture-level) models into timed automata (for formal verification) and Simulink models (for analysis, simulation, and code generation). Another possible solution is to rely on a common specification for defining different formalisms used for different purposes, and rightfully refining it when applied to a specific application domain. Recent researches attempt at bridging the gap between the three communities of model-based design, real-time analysis and model-driven development, for a better understanding of the ways in which new development flows can be constructed that go from system-level modeling to the correct and predictable generation of a distributed implementation. The following three chapters provide further progress in this direction.

The Chap. 4 (*Model-Driven Design of Software Defined Radio Applications based on UML*) describes a complete model-driven design methodology for software defined radio (SDR) applications. It includes a domain specific UML profile, DiplodocusDF, for SDR applications; a synthesis step that generates SystemC or C code from DiplodocusDF models; and a runtime environment for execution of the generated code. Finally, the proposed DiplodocusDF UML profile is supported by tools (TTool) for design exploration and formal verification at model level. The design methodology is illustrated with an example of a Welch periodogram detector.

The Chap. 5 (*On Integrating EAST-ADL and UPPAAL for Embedded System Architecture Verification*) is motivated by the need of a seamlessly integrated development environment that can ensure consistency between the constraints specified for different parts of a system. The chapter introduces a method to transform between EAST-ADL and timed automata, to facilitate the formal analysis of the consistency of EAST-ADL based timing constraint specifications. To verify the effectiveness of this method, a case study of a brake-by-wire system is used with analysis of the constraints specified on the periodicity and precedence of event occurrences, such as function triggers and data arrival on a port.

The Chap. 6 (*Schedulability Analysis at Early Design Stages with MARTE*) presents Optimum, a UML front-end for the designer that conforms to a formal model for schedulability analysis and supports the evaluation of different architecture candidates. It specifies a subset of MARTE concepts that restricts the use of some elements to express a precise semantics, but is sufficient for integrating schedulability analysis in specification, design and verification/validation stages of the software life cycle. The methodology has been successfully applied in the automotive domain in the context of two collaborative projects, the European INTERESTED project and the French national EDONA project.

1.4 Component-Based Design and Real-Time Components

Component-based or component-oriented development places emphasis on the capability of defining units of reuse at different levels of granularity and possibly at different levels in the refinement flow that can ease the task of building a correct complex system by leveraging a number of assets:

- A formal language for the definition of the structure and behavior of the component. The extent of this language can be quite broad. All the interaction points of the component with the external world need to be defined, whether in a data-oriented or service-oriented way. Similarly, the internal behavior must be formally specified and connected to the interface specification defining, when necessary, which sequences of data exchanges and service requests over the interface correspond to legal behaviors.
- The capability of analyzing and verifying the properties of the component if taken in isolation.
- The capability of guaranteeing the correct functioning of the component (the preservation of its properties) under any legal composition with other components (this of course entails the definition of a *legal composition*).
- The capability of analyzing properties of the composites starting from the properties of the components or, even better, of guaranteeing such properties *by construction*.

The specification of a component structure and behavior is often performed under the terms of a *contract* [55]. This definition, which originated the term *design-by contract* [56], has the merit of emphasizing the relationship between the component and the outside world, consisting of the other components and the environment that the system is controlling or in which it executes, and bringing to the foreground the necessary consideration that each component works under a set of *Assumptions* on the outside components and environment. If these assumptions are met, then the component guarantees the truth of a set of *Assertions*, which is the promise that the component developer makes on its side of the contract.

Building a full-fledged component specification language that satisfies the above requirements for an arbitrarily wide set of properties is unfortunately still utopistic, and most languages for the definition of components and the expression of compositionality rules focus on a specific set of requirements (or preferably assumptions/assertions).

Recently, time specifications have acquired a special importance, not only because of the growing use of models and electronics/software content for the development of time-critical functionality, but also because of the importance that an accurate timing analysis may have for the selection of the best execution architecture for a given set of functional components.

The Chap. 7 (*Early Time-Budgeting for Component-Based Embedded Control Systems*) focuses on the definition of a set of parametric specifications for the definition of the timed behavior of components. The time specification, formalized using Parametric Temporal Logic, can be used to expresses properties related to the worst-case response of component reactions. The compositionality of the PTL specifications allows to analyze system properties resulting from the composition of components in a large system architecture. The authors show how this analysis capability can be used not only for system analysis and verification, but also, in an architecture exploration process, to define the best decomposition and placement of functions.

The Chap. 8 (*Contract-Based Reasoning for Component Systems with Rich Interactions*) provides the formal definition of a language for the definition of components contracts. A rule unifying circular and non-circular assume-guarantee reasoning is proposed to check contract dominance which does not require the explicit composition of contracts. It is in particular useful to combine contract-based verification tools and corresponding results from two different component frameworks.

The Chap. 9 (*Extracting End-to-end Timing Models from Component-based Distributed Embedded Systems*) discusses a method for extracting system-level time assertions from the time specifications of distributed real-time components. It finds solutions to issues involved during model extraction, such as extraction of timing information from all nodes and networks in the system, and linking of trigger and data chains in distributed transactions.

1.5 Timing Analysis and Time-Based Synthesis

Embedded systems are very different from general purpose computing because of their dedication to control and because they interact with the physical environment. One of the most important constraints is that the time needed to perform a computattion or control task may be critical to the system correctness (this property defines the *hard* real-time systems). Starting from the 90s, the results of the real-time research community have been increasingly used in the industry for the development and analysis of operating systems scheduling policies and network medium access control protocols. The real-time systems community has traditionally considered tasks or jobs (from the operating system concept of thread) as the units for the analysis model. In hard real-time systems, the design space (or the feasibility region) must satisfy the schedulability constraints, requiring that tasks complete before their deadlines. In the design of these time-critical applications, schedulability analysis is used to define the feasibility region of tasks with deadlines, so that iterative analysis or optimization techniques can find the best design solution within the feasibility region.

Today, the increasing complexity and distribution of real-time embedded systems (e.g. around one hundred electronic control units, ten communication buses, and millions of lines of code in a modern automobile) often results in long design iterations to improve the design and fix errors, and ultimately sub-optimal solutions. The perspective of using timing analysis techniques has been largely changed: it needs to be put forward as much as possible in the design flow, from the analysis of a given configuration to the synthesis of an optimal design. Developers are today increasingly faced with design problems, including the optimal placement of functions, the optimal assignment of priorities (or time slots) to tasks and messages, and the optimal packing of communication signals in frames.

Practical design problems, are typically very complicated. For example, even if we only consider the problem of optimal priority assignment, there are many cases where finding the optimal solution is of exponential complexity (and deadline monotonic (DM) [57] (that is, executing first the most urgent tasks) is no longer guaranteed to be optimal):

- systems with non-preemptive scheduling [58] or in which tasks share resources [59];
- systems in which the designer seeks an optimal solution that maximizes the robustness (for example, the difference between the deadline and the worst-case response time) [60];
- systems in which preemption thresholds are used to optimize stack memory usage [61];
- systems that contain computing paths deployed on a distributed platform and constrained by end-to-end deadlines (like those dealt in the first chapter of the fourth part);
- systems with multiple control loops scheduled on the same processor and the objective function is the overall control performance [62, 63].

Because of the extremely large design space, a trial-and-error approach, in which a configuration is manually defined, analyzed for schedulability, and then possibly improved or fixed, is no longer practical. This calls for synthesis and optimization methods that can close the design loop automatically. The problem is of course also relevant for purely hardware (or programmable hardware) embedded designs, where the need for design synthesis has been established for quite some time now.

The Chap. 10 (*Distributed Priority Assignment in Real-Time Systems*) presents a recent advance on the problem of scheduling priority assignment for distributed real-time systems scheduled with preemptive static-priority under the consideration of end-to-end path latencies. The proposed algorithm relies on a distributed implementation of compositional performance analysis. It is implemented distributedly to reduce runtime overhead and to integrate with the admission control scheme. To prevent oscillation and help the distributed algorithm to converge, a time-discrete PID feedback control technique is proposed. The proposed algorithm can compete with state-of-the-art design time tools but only requires a fraction of the runtime.

The Chap. 11 (*Exploration of Distributed Automotive Systems using Compositional Timing Analysis*) presents a design space exploration approach for real-time automotive systems. First, it introduces the system model and a binary encoding to represent the design space. Then the chapter focuses on the timing analysis aspect (assuming the design choices have been made), in particular the issue of addressing cyclic dependencies in fixed-point iteration for timing analysis. The key idea is to build a dependency graph and leverage it for efficient analysis, and further improve the method by clustering local nodes and performing a hierarchical analysis.

The Chap. 12 (*Design and Evaluation of Future Ethernet AVB-based ECU Networks*) considers the design of automotive architectures based on the new high-speed communication media, Ethernet AVB, which is a light-weight Ethernet extension with prioritization and traffic shaping techniques to enhance its QoS aspects. The chapter presents a virtual prototyping approach with key features for the sake of design space exploration of extensive automotive architectures, such as static routing and stream reservation, fixed topology, and real-time applications. The proposed technique is an important step towards automatic optimization of architecture design in the presence of high-speed bus systems with enhanced QoS capabilities.

References

1. The MathWorks Inc.: Simulink User's Guide (2005). http://www.mathworks.com
2. Esterel Technologies: SCADE suite. http://www.esterel-technologies.com/products/scade-suite/
3. Andrade, H.A., Kovner, S.: Software synthesis from dataflow models for G and LabVIEW. In: Proceedings of the IEEE Asilomar Conference on Signals, Systems, and Computers, 1705–1709 (1998)
4. Modelica Association: Modelica and the modelica association. http://www.modelica.org/
5. Eker, J., Janneck, J.W., Lee, E.A., Liu, J., Liu, X., Ludvig, J., Neuendorffer, S., Sachs, S., Xiong, Y.: Taming heterogeneity-the Ptolemy approach. Proc. IEEE **91**(1), 127–144 (2003)
6. Davare, A., Densmore, D., Meyerowitz, T., Pinto, A., Sangiovanni-Vincentelli, A., Yang, G., Zeng, H., Zhu, Q.: A next-generation design framework for platform-based design. DVCon, In (2007)
7. Alur, R., Dill, D.L.: A theory of timed automata. Theor. Comput. Sci. 126(2), 183–235 (1994). http://dx.doi.org/10.1016/0304-3975(94)90010--8
8. Object Management Group: Omg model driven architecture. http://www.omg.org/mda/
9. Unified Modeling Language, 2.0. http://www.omg.org/uml
10. System Modeling Language. http://www.omg.org/sysml
11. Object Management Group: Concrete syntax for uml action language (action language for foundational uml - alf). http://www.omg.org/spec/ALF
12. Object Management Group: Semantics of a foundational subset for executable uml models (fuml). http://www.omg.org/spec/FUML
13. Object Management Group: UML profile for modeling and analysis of real-time and embedded systems (MARTE), version 1.1, formal/2011-06-02 (June 2011). http://www.omg.org/spec/MARTE/1.1/
14. Kienhuis, B., Deprettere, E.F., Wolf, P.v.d., Vissers, K.A.: A methodology to design programmable embedded systems–the y-chart approach. In: Embedded Processor Design Challenges: Systems, Architectures, Modeling, and Simulation–SAMOS, pp. 18–37. Springer-Verlag, London, UK, (2002)
15. Keutzer, K., Newton, A., Rabaey, J., Sangiovanni-Vincentelli, A.: System-level design: orthogonalization of concerns and platform-based design. Comput.-Aided Des. Integr. Circuits Syst., IEEE Trans. on **19**(12), 1523–1543 (2000)
16. Mukerji, J., Miller, J.: Overview and guide to omg's, architecture. http://www.omg.org/cgi-bin/doc?omg/03-06-01
17. Balarin, F., Watanabe, Y., Hsieh, H., Lavagno, L., Passerone, C., Sangiovanni-Vincentelli, A.L.: Metropolis: An integrated electronic system design environment. IEEE Comput. **36**(4), 45–52 (2003)
18. Object Management Group: Home page. http://www.omg.org/

19. T., B.: Current trends in the design of automotive electronic systems. Proceedings of the Design Automation and Test in Europe Conference (2001)
20. AUTOSAR consortium: automotive open system architecture. http://www.autosar.org
21. Edwards, S., Lavagno, L., Lee, E.A., Sangiovanni-Vincentelli, A.: Design of embedded systems: Formal models, validation and synthesis. Proc. IEEE **85**(3), 366–390 (1997)
22. Alur, R., Henzinger, T.A.: Logics and models of real time: A survey. Real-Time: Theory in Practice, REX Workshop, LNCS 600pp, 74–106 (1991)
23. Lee, E., Sangiovanni-Vincentelli, A.: A framework for comparing models of computation. Comput.-Aided Des. Integr. Circuits Syst., IEEE Trans. on **17**(12), 1217–1229 (1998)
24. Benveniste, A., Caspi, P., Edwards, S.A., Halbwachs, N., Guernic, P.L., Robert, Simone, D.: The synchronous languages 12 years later. In: Proceedings of The IEEE, pp. 64–83 (2003)
25. Berry, G., Gonthier, G.: The esterel synchronous programming language: design, semantics, implementation. Sci. Comput. Program. **19**(2), 87–152 (1992)
26. Boussinot, F., De Simone, R.: The esterel language. Proc. IEEE **79**(9), 1293–1304 (1991)
27. Caspi, P., Pilaud, D., Halbwachs, N., Plaice, J.A.: Lustre: a declarative language for real-time programming. In: Proceedings of the 14th ACM SIGACT-SIGPLAN symposium on Principles of programming languages, POPL '87, pp. 178–188. ACM (1987)
28. Halbwachs, N., Caspi, P., Raymond, P., Pilaud, D.: The synchronous dataflow programming language lustre. In: Proceedings of the IEEE, pp. 1305–1320 (1991)
29. Harel, D.: Statecharts: A visual formalism for complex systems. Sci. Comput. Program. **8**(3), 231–274 (1987)
30. Berthomieu, B., Diaz, M.: Modeling and verification of time dependent systems using time petri nets. Softw. Eng., IEEE Trans. **17**(3), 259–273 (1991)
31. Yovine, S.: Kronos: A verification tool for real-time systems. (kronos user's manual release 2.2). Int. J. Softw. Tools Technol. Transf. **1**, 123–133 (1997)
32. Behrmann, G., David, A., Larsen, K.G.: A tutorial on uppaal. In: M. Bernardo, F. Corradini (eds.) Formal methods for the design of real-time systems: 4th International School on Formal Methods for the Design of Computer, Communication, and Software Systems, SFM-RT 2004, no. 3185 in LNCS, pp. 200–236. Springer-Verlag (2004)
33. Wozniak, E., Mraidha, C., Gerard, S., Terrier, F.: A guidance framework for the generation of implementation models in the automotive domain. In: Software Engineering and Advanced Applications (SEAA), 2011 37th EUROMICRO Conference on, pp. 468–476 (2011)
34. Koudri, A., Cuccuru, A., Gerard, S., Terrier, F.: Designing heterogeneous component based systems: evaluation of MARTE standard and enhancement proposal. In: Model Driven Engineering Languages and Systems, pp. 243–257 (2011)
35. EAST-ADL Overview. http://www.atesst.org/home/liblocal/docs/ConceptPresentations/01_EAST-ADL_OverviewandStructure.pdf
36. Department of Defense: DoD architecture framework v2.02. http://dodcio.defense.gov/Portals/0/Documents/DODAF/DoDAF_v2-02_web.pdf
37. Vanderperren, Y., Dehaene, W.: From uml/sysml to matlab/simulink: current state and future perspectives. In: Proceedings of the conference on Design, automation and test in Europe, DATE '06. Leuven, Belgium (2006)
38. Berry, G., Gonthier, G.: The synchronous programming language esterel: Design, semantics, implementation. Sci. Comput. Program. **19**(2), 87–152 (1992)
39. Ledeczi, A., Maroti, M., Bakay, A., Karsai, G., Garrett, J., Thomason, C., Nordstrom, G., Sprinkle, J., Volgyesi, P.: The generic modeling environment. Workshop on Intelligent Signal Processing, In (2001)
40. Sangiovanni-Vincentelli, A., Shukla, S., Sztipanovits, J., Yang, G., Mathaikutty, D.: Metamodeling: An emerging representation paradigm for system-level design. Special Section on Meta-Modeling, IEEE Des. Test **26**(3), 54–69 (2009)
41. Alfaro, L., Henzinger, T.A.: Interface automata. In: Proceedings of ESEC/SIGSOFT FSE'01, pp. 109–120. ACM Press (2001)
42. Lee, E., Sangiovanni-Vincentelli, A.: A unified framework for comparing models of computation. In. IEEE Trans. Comput. Aided Des. Integr. Circuits Syst. **17**, 1217–1229 (1998)

43. Liu, C.L., Layland, J.W.: Scheduling algorithms for multiprogramming in a hard-real-time environment. J. ACM **20**, 46–61 (1973)
44. Lehoczky, J.P., Sha, L., Ding, Y.: The rate-monotonic scheduling algorithm: exact characterization and average case behavior. In: Proceedings of the 10th IEEE RTSS, pp. 166–171. Santa Monica, CA USA (1989)
45. Benveniste, A., Caillaud, B., Carloni, L.P., Caspi, P., Sangiovanni-Vincentelli, A.L., Tripakis, S.: Communication by sampling in time-sensitive distributed systems. In: EMSOFT, pp. 152–160 (2006)
46. Zheng, W., Natale, M.D., Pinello, C., Giusto, P., Sangiovanni-Vincentelli, A.: Synthesis of task and message activation models in real-time distributed automotive systems. In: DATE'07: Proceedings of the Design, Automation and Test in Europe Conference. Nice, France (2007)
47. Davare, A., Zhu, Q., Natale, M.D., Pinello, C., Kanajan, S., Sangiovanni-Vincentelli, A.: Period optimization for hard real-time distributed automotive systems. In: DAC '07: Proceedings of the 44th annual conference on Design automation, pp. 278–283. ACM, New York, NY, USA (2007) http://doi.acm.org/10.1145/1278480.1278553
48. Tindell, K.W.: Holistic schedulability analysis for distributed hard real-time systems. Tech. Rep. YCS 197, Department of Computer Science, University of York (1993)
49. Palencia, J., Harbour, M.G.: Schedulability analysis for tasks with static and dynamic offsets. In: 19th IEEE Real-Time Systems Symposium. Madrid, Spain (1998)
50. Di Natale, M., Zheng, W., Pinello, C., Giusto, P., Sangiovanni Vincentelli, A.: Optimizing end-to-end latencies by adaptation of the activation events in distributed automotive systems. In: Proceedings of the IEEE Real-Time Application Symposium. Bellevue, WA (2007)
51. Hamann, A., Henia, R., Jerzak, M., Racu, R., Richter, K., Ernst, R.: SymTA/S symbolic timing analysis for systems. available at http://www.symta.org (2004)
52. RTCA: Do-178b: Software considerations in airborne systems and equipment certification. http://www.rtca.org/
53. ISO: Iso 26262 road vehicles-functional safety. http://www.iso.org/
54. Benveniste, A., Caspi, P., Guernic, P.L., Marchand, H., Talpin, J.P., Tripakis, S.: A protocol for loosely time-triggered architectures. In: Proceedings of the Second International Conference on Embedded Software, EMSOFT '02, pp. 252–265. Springer-Verlag, London, UK, (2002)
55. Sangiovanni-Vincentelli, A., Damm, W., Passerone, R.: Taming Dr. Frankenstein: Contract-based design for cyber-physical systems. Eur. J. Control **18**(3), 217–238 (2012). doi:10.3166/EJC.18.217-238
56. B., M.: An overview of Eiffel. In: The Handbook of Programming Languages, vol. 1, Object-Oriented Languages, ed. Peter H. Salus, Macmillan Technical Publishing (1998)
57. Audsley, N.C., Burns, A., Davis, R., Tindell, K.W., Wellings, A.J.: Fixed priority pre-emptive scheduling: an historical prespective. Real-Time Syst. **8**, 173–198 (1995)
58. George, L., Rivierre, N., Spuri, M.: Preemptive and Non-Preemptive Real-Time UniProcessor Scheduling. Research Report RR-2966, INRIA (1996)
59. Mok, A.K.: Fundamental design problems of distributed systems for the hard-real-time environment. Ph.d. thesis, Cambridge, MA, USA (1983)
60. Hamann, A., Racu, R., Ernst, R.: Multi-dimensional robustness optimization in heterogeneous distributed embedded systems. In: Proceedings of the 13th IEEE Real Time and Embedded Technology and Applications Symposium, RTAS '07, pp. 269–280. IEEE Computer Society, Washington, DC, USA (2007)
61. Ghattas, R., Dean, A.G.: Preemption threshold scheduling: Stack optimality, enhancements and analysis. In: Proceedings of the 13th IEEE Real Time and Embedded Technology and Applications Symposium, RTAS '07, pp. 147–157. IEEE Computer Society, Washington, DC, USA (2007)
62. Cervin, A., Henriksson, D., Lincoln, B., Eker, J., Arzen, K.: How does control timing affect performance? analysis and simulation of timing using jitterbug and truetime. Control Syst., IEEE **23**(3), 16–30 (2003)
63. Velasco, M., Martí, P., Bini, E.: Control-driven tasks: Modeling and analysis. In: Proceedings of the 2008 Real-Time Systems Symposium, RTSS '08, pp. 280–290. IEEE Computer Society, Washington, DC, USA (2008)

Part I
Model-Based Design and Synthesis

Chapter 2
Modeling, Analysis, and Implementation of Streaming Applications for Hardware Targets

Kaushik Ravindran, Arkadeb Ghosal, Rhishikesh Limaye, Douglas Kim, Hugo Andrade, Jeff Correll, Jacob Kornerup, Ian Wong, Gerald Wang, Guang Yang, Amal Ekbal, Mike Trimborn, Ankita Prasad and Trung N. Tran

Abstract Application advances in the signal processing and communications domains are marked by an increasing demand for better performance and faster time to market. This has motivated model-based approaches to design and deploy

K. Ravindran (✉) · A. Ghosal · R. Limaye · D. Kim · H. Andrade · J. Correll · J. Kornerup · I. Wong · G. Wang · G. Yang · A. Ekbal · M. Trimborn · A. Prasad · T. N. Tran
National Instruments Corporation, Berkeley, CA, USA
e-mail: kaushik.ravindran@ni.com

A. Ghosal
e-mail: arkadeb.ghosal@ni.com

R. Limaye
e-mail: rhishikesh.limaye@ni.com

D. Kim
e-mail: douglas.kim@ni.com

H. Andrade
e-mail: hugo.andrade@ni.com

J. Correll
e-mail: jeff.correll@ni.com

J. Kornerup
e-mail: jacob.kornerup@ni.com

I. Wong
e-mail: ian.wong@ni.com

G. Wang
e-mail: gerald.wang@ni.com

G. Yang
e-mail: guang.yang@ni.com

A. Ekbal
e-mail: amalekbal@ni.com

M. Trimborn
e-mail: mike.trimborn@ni.com

A. Sangiovanni-Vincentelli et al. (eds.), *Embedded Systems Development*, Embedded Systems 20, DOI: 10.1007/978-1-4614-3879-3_2,

such applications productively across diverse target platforms. Dataflow models are effective in capturing these applications that are real-time, multi-rate, and streaming in nature. These models facilitate static analysis of key execution properties like buffer sizes and throughput. There are established tools to generate implementations of these models in software for processor targets. However, prototyping and deployment on hardware targets, in particular reconfigurable hardware such as FPGAs, are critical to the development of new applications. FPGAs are increasingly used in computing platforms for high performance streaming applications. They also facilitate integration with real physical I/O by providing tight timing control and allow the flexibility to adapt to new interface standards. Existing tools for hardware implementation from dataflow models are limited in their ability to combine efficient synthesis and I/O integration and deliver realistic system deployments. To close this gap, we present the LabVIEW DSP Design Module from National Instruments, a framework to specify, analyze, and implement streaming applications on hardware targets. DSP Design Module encourages a model-based design approach starting from streaming dataflow models. The back-end supports static analysis of execution properties and generates implementations for FPGAs. It also includes an extensive library of hardware actors and eases third-party IP integration. Overall, DSP Design Module is an unified design-to-deployment framework that translates high-level algorithmic specifications to efficient hardware, enables design space exploration, and generates realistic system deployments. In this chapter, we illustrate the modeling, analysis, and implementation capabilities of DSP Design Module. We then present a case study to show its viability as a model-based design framework for next generation signal processing and communications systems.

2.1 Introduction

Dataflow models are widely used to specify, analyze, and implement multi-rate computations that operate on streams of data. Static Dataflow (SDF)[1] is a well known model of computation for describing signal processing applications [1]. An SDF model is a graph of computation actors connected by channels that carry data tokens. The semantics require the number of tokens consumed and produced by an actor per firing be fixed and pre-specified. This guarantees static analyzability of key execution properties, such as deadlock-free operation and bounded memory requirements [2].

A. Prasad
e-mail: ankita.prasad@ni.com

T. N. Tran
e-mail: trung.tran@ni.com

[1] SDF was called *Synchronous* Dataflow in the original works that introduced the model [1, 2]. But the model is fundamentally asynchronous, since actors can fire independently and asynchronously. For this reason, and in order not to confuse SDF with truly synchronous models such as synchronous FSMs, we prefer the term *Static* Dataflow.

The expressiveness of dataflow models in naturally capturing streaming applications, coupled with formal analyzability properties, has made them popular in the domains of multimedia, signal processing, and communications. These high level abstractions are the starting points for model-based design approaches that enable productive design, fast analysis, and efficient correct-by-construction implementations. Ptolemy II [3], Grape-II [4], LabVIEW [5], and Simulink [6] are examples of successful tools built on the principles of model-based design from dataflow models. These tools provide a productive way to translate domain specific applications to efficient implementations.

Most of these tools predominantly deliver software implementations for general purpose and embedded processor targets. However, ever-increasing demands on performance and throughput of new applications and standards have motivated prototyping and deployment on hardware targets, in particular reconfigurable hardware such as Field Programmable Gate Arrays (FPGAs). FPGAs are integral components of modern computing platforms for high performance signal processing. The configurability of FPGAs and constraints of hardware design bring unique implementation challenges and performance-resource trade-offs. FPGAs permit a range of implementation topologies of varying degrees of parallelism and communication schemes. Another requirement for hardware design is the integration of pre-created configurable intellectual property (IP) blocks. Actor models must capture relevant variations in data access patterns and execution characteristics of IP configurations.

Common software synthesis techniques for dataflow models are not immediately applicable to hardware synthesis. Related tools from prior works either make restrictive assumptions about the dataflow models to simplify hardware generation, or do not provide a complete path to implementation on realistic targets supporting established design methodologies and IP libraries. They do not fully address the complexities of connecting hardware actors and IP with differing interfaces. The tools are limited in their ability to deliver realistic system deployments, combining efficient hardware synthesis and physical I/O integration.

In this work, we advance a design framework to enable the design and deployment of streaming applications on FPGA targets. The objective is to empower application and domain experts without specialized knowledge in hardware design to create efficient implementations on FPGA targets. Founded on model-based design principles, the framework emphasizes six key elements to address the hardware design challenge:

- An appropriate model of computation to capture high-level intent.
- An IP library for general computing and domain specific functions.
- A graphical interface that incorporates latest interactive technologies.
- Measurement and control I/O to design and test such systems.
- Static analysis and evaluation of key trade-offs at design time.
- Efficient deployments on high performance hardware platforms.

We combine these elements in the National Instruments (NI) LabVIEW DSP Design Module, a framework for hardware-oriented specification, analysis, and implementation of streaming dataflow models. The framework enables DSP domain experts to express complex applications and performance requirements in an algorithmic manner and automatically generate efficient hardware implementations. The main components of DSP Design Module are: (a) a graphical specification language to design streaming applications, (b) an analysis engine to validate the model, select buffer sizes and optimize resource utilization to meet throughput constraints, and perform other pertinent optimizations, and (c) implementation support to generate an efficient hardware design and deploy it on Xilinx FPGAs.

The specification is based on the Static Dataflow (SDF) model of computation with extensions to support cyclo-static data rates and parameterization. Prior studies on modeling requirements for wireless communications applications endorse that this model is sufficiently expressive in capturing relevant execution and performance characteristics [7, 8]. DSP Design Module provides an extensive library of math and signal processing functions that harness the resource elements on the FPGA. It also facilitates integration of custom-designed hardware blocks and third-party IP into the design. The back-end eases exploration of design trade-offs and translates a high level algorithmic specification to an efficient hardware implementation.

DSP Design Module enables application domain experts implement complex streaming applications on FPGAs. The hardware implementations are currently targeted to Xilinx Virtex-5 FPGAs on the National Instruments PXIe platform. This is an important distinction from prior works which stop with RTL synthesis from high level models. In contrast, DSP Design Module is integrated with LabVIEW, a commercial framework for deploying systems that combine processors, FPGAs, real-time controllers, and analog and digital I/O. In particular, the tight contract-based integration between DSP Design Module and LabVIEW FPGA enables orthogonalization of design. One designer can focus on the details of integration to I/O via LabVIEW FPGA and control the tight timing requirements of hardware interfaces at the cycle level. Another designer can specify the I/O requirements as equivalent port constraints on a DSP Design Module algorithm description, so that the tool can generate code that satisfies these constraints. DSP Design Module hence delivers a unified design-to-deployment framework for streaming applications.

In this chapter, we highlight salient features of the NI LabVIEW DSP Design Module and illustrate a design flow to implement streaming applications. We then present a case study on the design and implementation of a real-time single antenna Orthogonal Frequency Division Multiplexing (OFDM) wireless communication link transmitter and receiver. The OFDM is part of the Long Term Evolution (LTE) mobile networking standard [9]. The system is implemented on the National Instruments PXIe platform equipped with Xilinx Virtex-5 FPGAs, real-time controllers, and baseband and RF transceivers. We demonstrate the design of key application components in DSP Design Module and show how the analysis capabilities of the tool help the designer make key performance and resource trade-offs. We finally show the deployment of the system on the PXIe platform. The study illustrates how DSP Design Module serves as a model-based design framework that translates high level streaming models to efficient hardware implementations.

2.2 Related Work

Synthesis flows from Register Transfer Level (RTL) logic and behavioral languages (typically C, C++, or SystemC) for hardware targets has been a popular topic of several studies. However, there is limited prior art on hardware generation from non-conventional high level models like dataflow. Ptolemy II is a prominent academic framework for graphical specification and analysis of multiple models of computation [3]. Grape-II facilitates emulation of SDF models on FPGAs and explores memory minimization techniques [4]. While these tools provide some support for RTL generation from restricted models, the focus is more on proof-of-concept and less on optimized hardware implementation.

The work of Edwards and Green also confirms these limitations [10]. They explicitly address hardware generation and evaluate architectures for implementation of SDF models. They also discuss strategies to optimize function parallelism and token buffering on FPGAs. Horstmannshoff et al. [11] and Jung et al. [12] also discuss hardware synthesis techniques from SDF models. They develop analysis methods for resource allocation and schedule computation suitable for hardware. However, these prior works do not consolidate the proposed techniques into a comprehensive design framework that enables implementation on realistic targets. They do not support established synthesis methodologies and IP libraries. Actor definition and IP integration are practical challenges in hardware design, yet these works do not address them adequately.

On the commercial front, LabVIEW FPGA from National Instruments is a popular tool that supports FPGA deployment from dataflow models [13]. However, LabVIEW FPGA only supports the Homogeneous Static Dataflow (HSDF) model of computation, which does not natively capture streaming multi-rate computations. Furthermore, it requires explicit specification of cycle-level behavior for generating efficient hardware. System Generator from Xilinx is another related offering that supports FPGA implementations of synchronous reactive and discrete time models of computation [14]. However, these models are not suitable for data driven streaming specifications that are better expressed in dataflow. SystemVue ESL from Agilent supports more expressive dataflow models and provides libraries and analysis tools for the RF and DSP domains [15]. However, it primarily serves as an exploration and simulation environment, and does not offer a path to implementation in hardware. To our knowledge, there is no integrated tool that addresses the challenges in translating dataflow models to implementations on practical hardware targets.

The closest effort in synthesizing hardware from dataflow programs is the Open Dataflow framework [16]. The CAL actor language in Open Dataflow is an important step in formalizing actor and interface definitions. It has been adopted by the MPEG Video Coding group to develop codecs for future standards [17]. CAL builds on the Dynamic Dataflow model of computation, which allows specifications with variable data rates and non-deterministic behaviors. But this model is undecidable and cannot be subject to static analysis. There have been recent efforts to detect analyzable regions in a CAL model that adhere to restricted models of computation [18].

However, the language does not constrain the designer from introducing structures that might prohibit analysis. In contrast, the DSP Design Module model of computation is intentionally restrictive in order to enable efficient analysis of deadlock-free execution, buffer size, and throughput. While the model may not be as expressive as CAL, it is nevertheless applicable to a range of interesting signal processing applications. Also, CAL is a textual specification language, whereas DSP Design Module is a graphical design environment. We believe a graphical framework built on analyzable models provides an intuitive design approach for domain experts not specialized in hardware design to create efficient system deployments.

In summary, DSP Design Module is an attempt to integrate the respective strengths of the previously discussed tools into a unified design-to-deployment framework for streaming applications on hardware targets. The graphical design environment is intended for algorithm designers who are generally not experts in hardware design. The framework supports analysis capabilities relevant to hardware implementation and includes an extensive library of common math, signal processing, and communications functions. It also enables integration of IPs from native and third-party libraries, like the Xilinx CoreGen library [19], which are essential to practical efficient hardware design. Finally, it delivers implementations on realistic hardware targets that can be combined with other processors, memories, and physical I/O to build high performance embedded systems.

2.3 DSP Design Module: Models and Analysis

The DSP Design Module captures specifications based on the Static Dataflow (SDF) model of computation, with extensions for cyclo-static data rates and parameterization. This section presents the relevant characteristics of these models and illustrates their suitability for specifying signal processing applications.

2.3.1 Static Dataflow

A dataflow model consists of a set of actors (which represent computational units) inter-connected via channels (which communicate data, abstracted as tokens, between actors). In the *Static Dataflow* (SDF) model of computation, at each firing, an actor consumes a fixed number of tokens from each input channel, and produces a fixed number of tokens on each output channel . A channel stores the generated tokens until an actor consumes the tokens. We illustrate the SDF model and relevant properties using the model shown in Fig. 2.1. The model is composed of two actors: *distribute stream*, and *interleave stream*, henceforth referred to as *distribute* and *interleave* actors. The *distribute* actor splits an incoming stream of data into two output streams, while the *interleave* actor merges two incoming streams into one outgoing stream. The *distribute* actor consumes 2 tokens from the incoming channel and produces 1 token on each of the two output channels; the effect of the actor

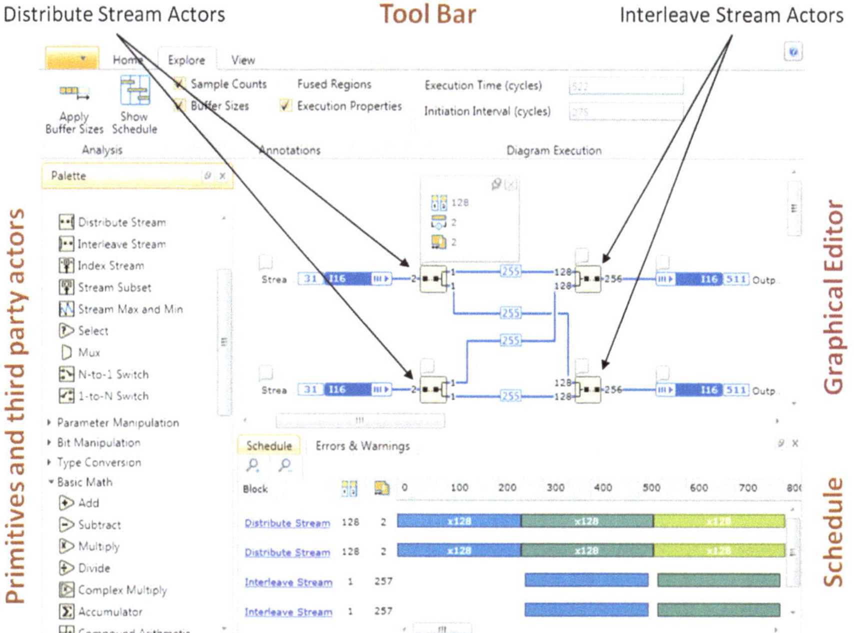

Fig. 2.1 This is the LabVIEW DSP design module graphical user interface in which an SDF model is captured. At the *top* is the tool bar, where the user can set platform characteristics and select different analysis functions. At the *left* is the tool library with different math and signal processing primitives and third party actors. In the *middle* is the graphical canvas where the SDF model is specified. The *lower part* shows the execution schedule along with any errors and warnings

processing is that the first (resp. second) outgoing stream consists of tokens from odd (resp. even) numbered positions of the incoming stream. The *interleave* actor consumes 128 tokens from each of the two incoming channels and produces 256 tokens on the outgoing channel. The actor interleaves the tokens such that the outgoing stream is a concatenation of blocks of 128 tokens taken in alternating fashion from the two input streams. The model has two instances of the *distribute* and *interleave* actors, with each instance of the *distribute* actors connected to both instances of the *interleave* actors. The actors do not change the tokens; they only rearrange the tokens at the outputs so that outgoing streams consist of a different sequence of tokens from the incoming streams.

Each actor is associated with an *execution time* and an *initiation interval*. Execution time is the time (in clock cycles) that the actor needs to process inputs, perform computation, and generate outputs. Initiation interval is the minimum time (in clock cycles) between consecutive firings of an actor. If initiation interval is less than execution time for an actor, the actor may fire in an overlapping (pipelined) fashion. In the above example, the execution time and the initiation interval of the *distribute* actor are 2 cycles each; the execution time and the initiation interval of the *interleave* actor are 257 cycles each.

2.3.2 SDF: Properties and Analysis

The SDF model of computation permits decidability of key properties of streaming applications, such as bounded memory and deadlock free execution [1]. A model is bounded if it can be executed without termination using buffers with finite capacity. An execution of an SDF model can achieve periodic behavior because the token consumption and production rates are constant. In SDF models, bounded execution is verified by proving that the model is *sample rate consistent* [2]. An SDF model is sample rate consistent if there exists a fixed non-zero number of firings for each actor such that executing these firings reverts the model to its original state. A vector denoting this number for each actor is called the *repetitions vector*. Consider a channel between any *distribute* and *interleave* actors in Fig. 2.1. Given that the *distribute* actor produces 1 token and the *interleave* actor consumes 128 tokens, the *distribute* actor needs to fire 128 times more often than the *interleave* actor in each round of firing. Hence, the repetitions count for the *distribute* actors is 128, while that for the *interleave* actors is 1. These repetition counts ensure that the number of tokens produced over a channel is equal to the number of tokens consumed. This in turn guarantees that any non-terminating periodic execution of the SDF model can be implemented with buffers that have bounded capacity.

In addition to boundedness, deadlock-free operation is essential for most streaming applications. Dataflow models that are bounded and deadlock-free are viable for implementation in hardware and software. An SDF model is *deadlock-free* if at any point in its execution, there is at least one actor that can fire. For a sample rate consistent SDF model, it suffices to verify that a single *iteration* (in which each actor fires as many times as specified in the repetitions vector) of the model can be completely executed. The solution is to compute a *self timed schedule* (in which an actor fires as soon as all input tokens are available) for one iteration [2]. As the model is consistent, this schedule can be executed infinitely often. This computation assumes that channels in the SDF model are implemented as FIFO buffers with infinite capacity. Nevertheless, a bound on the buffer size can be encoded as a back edge in the SDF model with initial tokens corresponding to the size of the buffer [20]. The model in Fig. 2.1 is bounded and deadlock free.

Important performance metrics for real-time streaming applications are throughput (i.e., rate of tokens produced by an actor) and memory (i.e., buffer sizes to implement channels). Each actor needs to wait until the respective producing actor has generated the required number of tokens, which imposes restrictions on the minimum buffer sizes required to implement the channels. For the example model in Fig. 2.1, the *interleave* actor needs 128 tokens per firing, whereas the *distribute* actor produces 1 token per firing, i.e., one firing of the *interleave* actor depends on 128 firings of the *distribute* actor. Hence any channel between the two actors must accommodate at least 128 tokens. Static analysis can also determine the buffer sizes of the channels needed to meet the throughput requirements of the model. Buffer sizes are dictated by the data rates on the channels, the throughput requirements, and the clock rate at which the model is executed. Throughput is usually specified in engineering units such as Mega-Samples per second (MSps). Setting the throughput at the outputs as 10 MSps for a clock rate of 40 MHz in the example model, the buffer size required for each channel is 255; the buffer size can achieve a maximum throughput of 37.23 MSps for the clock rate of 40 MHz. The *schedule* (i.e., order of firings) of the actors captures how the actors are executed in time. The schedule for the example (shown at the bottom of Fig. 2.1) shows that the *distribute* and *interleave* actors fire in parallel, with the *distribute* actors firing 128 times more often than the *interleave* actors.

2.3.3 Extensions for Cyclo-Static Data Rates and Parameterization

The SDF model of computation accurately captures fixed-length computations on streams. But there are some actors that follow a pre-specified cyclic pattern in the number of tokens processed. For such computations, the *Cyclo-Static Dataflow* (CSDF) [21] model of computation generalizes SDF by allowing the number of tokens consumed or produced by an actor to vary according to a fixed cyclic pattern. Each firing of a CSDF actor corresponds to a phase of the cyclic pattern. In Fig. 2.1, if the input token count of the *interleave* actors is replaced by a cyclic pattern (64, 128, 256), then the result is a CSDF model that interleaves incoming streams following a deterministic cyclic pattern of input token counts. As the cyclic pattern is fixed and known at design time, all static analysis properties of SDF are also applicable [22]. We refer to [20, 23] for details on analysis algorithms to compute buffer sizes and throughput.

SDF and CSDF are static in nature. However, some applications may require that the number of tokens processed by an actor vary at run-time. For example, consider a variant of the *interleave* actor that can be configured to consume 64, 128, or 256 tokens on its input channels per firing. The token consumption count on its input channels is no longer a fixed value; it is denoted by a parameter N, where N is from the set {64, 128, 256}. Note that this behavior is different from the CSDF actor discussed earlier. In successive firings, the CSDF actor consumes {64, 128, 256} in a cyclical pattern. In contrast, the parameterized actor consumes N tokens per firing, but the value the parameter is determined by external input and need not follow a pre-specified pattern. The *Parameterized Static Dataflow* (PSDF) model of computation is an extension to SDF that allows such parameterized actors [24].

The PSDF model can be viewed as a composition of several SDF models. At any point in execution, the behavior of a PSDF model corresponds to the underlying SDF model with all parameters fixed to a specific value. However, if the parameter values are allowed to change in any arbitrary manner, the problem of verifying whether the PSDF model is bounded and deadlock-free becomes undecidable. One useful restriction to facilitate static analysis is to allow changes in parameter values to take effect only at iteration boundaries. This restriction ensures that a PSDF model can be statically analyzed to verify bounded deadlock-free execution by verifying that the individual SDF models corresponding to different combinations of parameter values are bounded and deadlock-free. Subsequent analysis of throughput and buffer sizes can be computed as the worst case from among all combinations of parameter values. The CSDF model can similarly be parameterized to form the *Parameterized Cyclo-Static Dataflow* (PCSDF) model.

2.4 DSP Design Module: Implementation Flow

DSP Design Module is a graphical environment tool which allows user to specify streaming applications, explore optimizations, and generate synthesizable FPGA designs. Fig. 2.2 summarizes the design and implementation flow in DSP Design Module. In this section we discuss this flow using an OFDM transmitter application as a driving example.

2.4.1 Design Environment

The user works in a graphical environment as shown in Fig. 2.1. The starting point is the *Application*, e.g. a DSP algorithm, which the user starts drawing by selecting actors from the *Actor Library* and placing them on the editor canvas. This begins the

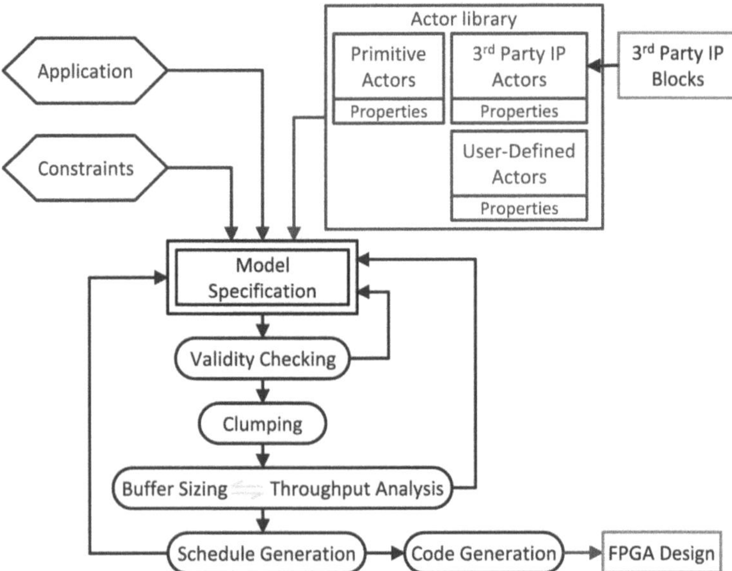

Fig. 2.2 Design and implementation flow in DSP design module

Model Specification step. The actor library consists of a rich set of primitive actors, including computational actors (add, multiply, square root, sine, log etc.), stream manipulation actors (upsample, downsample, distribute stream, interleave stream, etc.), third-party actors (e.g. FFT and FIR blocks from Xilinx Coregen [19]), and user-defined actors that are either specified in the LabVIEW programming language or previously defined models constructed using DSP Design Module. This reuse of actors allows for hierarchical composition of designs within the tool.

Figure 2.3 shows the SDF model for the OFDM Transmitter. The basic building blocks of the transmitter consist of data and reference symbol interleaving and zero padding the signal, followed by FFT processing and sample rate conversion. The input signals are first interleaved and distributed followed by zero padding. The streams are then processed by an FFT actor, and subsequently down sampled using two FIR filters. The FFT (Xilinx 7.0) and FIR (Xilinx 5.0) filters are third party actors. The other actors are primitives of the DSP Design Module. The user can also define actors using DSP Design Module. Figure 2.4 shows the same model where the computations for zero padding (performed by the distribute stream, the interleave stream, and the split number actors in Fig. 2.3) are grouped under an actor named ZeroPad (the first actor from left following the input ports).

The user describes the models by connecting the actors, and optionally configures their properties; e.g., FIR filter actors can be configured for different downsampling/up-sampling scenarios. Configurable properties of an actor include the data types and the number of tokens for the input and output channels. For some actors, throughput, pipeline depth, resource usage, or other implementation-specific options can also be configured by the user. The actor library includes cycle-accurate characteristics for each actor configuration, including initiation interval and execution time.

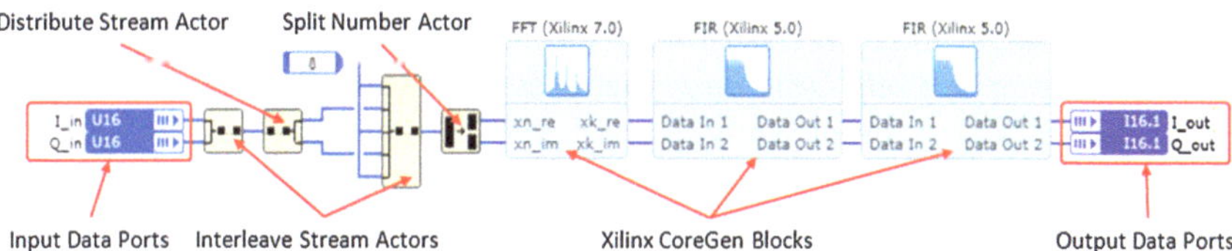

Fig. 2.3 SDF model of the OFDM transmitter

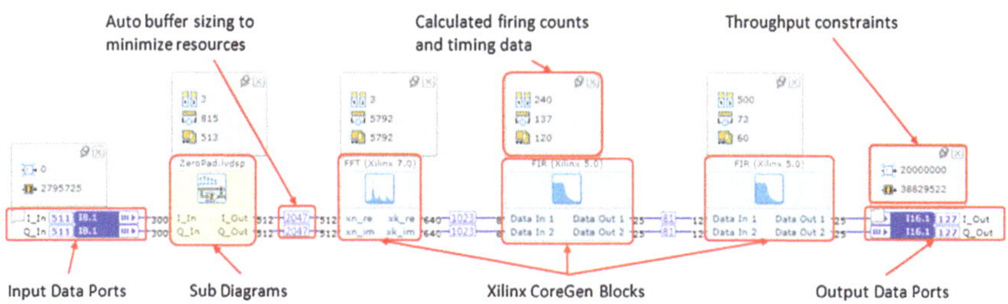

Fig. 2.4 Analysis results for the SDF model of the OFDM transmitter

In the OFDM example, the first FIR filter is configured with a user-defined coefficient file, and the interpolation and decimation factors are set to 25 and 8, respectively. The second FIR filter is configured with another user defined coefficient file, and the interpolation and decimation factors are set to 25 and 12, respectively.

The second input from the user is the *Constraints* for hardware generation. This includes minimum throughput requirements and target clock frequency. Throughputs can be associated with input/output ports or internal channels of the design, and are specified in engineering units, such as Mega-Samples per second (MSps). The model in Fig. 2.4 sets the minimum required throughput at the output ports to be 20 MSps and the target frequency to be 40 MHz.

The tool performs several types of analysis on the design in the background while the user is constructing it, with immediate feedback on the current state of the design. *Validity Checking* verifies bounded and deadlock-free execution of the model. It also performs automatic type checking and type propagation across the design. Errors or warnings are immediately annotated on the offending nodes on the canvas and reported under the Errors and Warning tab in the tool. On a valid design, the tool performs *Clumping* to identify homogeneous and simple regions that fit specialized implementation schemes (see Sect. 2.4.3). *Buffer Sizing* and *Throughput Analysis* are then performed on the design. This determines the buffer sizes required on the channels to satisfy user constraints such as minimum throughput. If the constraints cannot be met, the tool reports errors. *Schedule Generation* establishes a valid, cycle-accurate schedule (Fig. 2.5) for the design, given the determined buffer sizes and clumped regions. Thus the tool provides instant feedback on the run-time behavior of the design, including the achievable throughput.

The user can simulate the functional behavior on the development platform before invoking the hardware implementation stage. As a part of the simulation, the user can specify stimulus data and add graphical displays to the design to visualize the response on output ports or on any wire in the design.

The final step is *Code Generation* that uses the results of analysis to emit an FPGA design in the form of synthesizable LabVIEW files. The tool can also generate a synthesizable testbench that allows the user to stimulate the design from the development computer and compare the response to validated signals. The testbench includes the necessary code for DMA communication between the FPGA device and the development computer. The LabVIEW files can be used to generate a bitfile used to implement the design on Xilinx FPGA devices or for timing-accurate hardware simulation. Currently the tool supports targeting Virtex 5 devices from Xilinx.

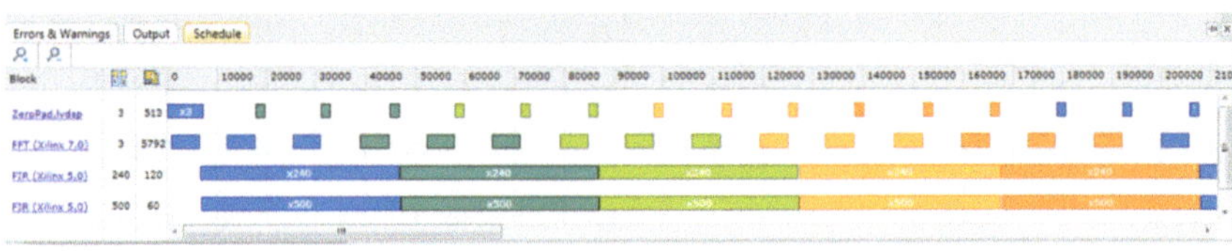

Fig. 2.5 Schedule for the SDF model of the OFDM transmitter

2.4.2 Implementation Strategy

DSP Design Module uses a FIFO-based, *self-timed* implementation strategy to realize the designs on FPGA fabrics [23]. In the FIFO-based strategy every channel in a model is conceptually mapped to a hardware FIFO of appropriate size and every actor is mapped to a dedicated hardware block that implements its functionality. There is no resource sharing among two different channels or two different actors in the current state of the tool, but the model does not preclude this. In the self-timed implementation strategy every actor is fired whenever it has a sufficient number of tokens on each of its input channels, sufficient number of vacancies on each of its output channels, and the initiation interval of the previous firing has expired. This evaluation is done for an actor on every clock cycle, and is independent of the state of the other actors in the model. As a consequence, there is no global scheduling logic in this implementation strategy, reducing the complexity of the controller for each actor in the design.

2.4.3 Glue Design and IP Integration

The FIFO-based, self-timed implementation strategy surrounds each actor with a harness that provides a FIFO interface to realize the SDF model and its extensions discussed in Sect. 2.3. The generated code for an actor presents a standardized interface to the harness, with designated lines for data and handshaking signals. A standardized harness simplifies *glue design*, i.e., the design of communication and control logic to stich hardware blocks together. Further, a standardized harness provides a mechanism to integrate native and third-party IP. The IP interfaces can be adapted to the harness template, which facilitates the generation of the *glue* to connect them to actors in the model.

A faithful realization of the SDF model of computation requires extra resources for the harness logic and the FIFOs on each channel. However, in the synthesized design this overhead can be significant compared to the resource usage of the actors themselves. To reduce this overhead the tool applies a series of *clumping* transformations to reduce the number of harnesses and FIFOs in the design. These transformations preserve the observable flow of tokens on the input and output ports, while preserving or increasing throughput. A clump is either a single actor or a collection of neighboring actors connected using direct locally synchronized connections without a FIFO-based harness interface. A clump exports a standard actor interface at its boundary. Hence it is a candidate for further clumping with other actors or clumped regions.

The supported clumping transformations are *Simple*, *Combinational*, and *Homogeneous* clumps. In a simple clump, design constants and downstream branches are merged with connected actors, resulting in a clump where the constant inputs are replaced with directly wired constants and the branches translate directly to branched output wires. Combinational clumps are constructed explicitly in the tool when the user selects connected actors in the design that can complete their entire execution in a single cycle, and designates them as executing jointly in a single cycle. The result is a clump with direct wire connections between the synthesized actors. Homogeneous clumps are regions where all actors are homogeneous, i.e., all the input and output counts are one. These actors are connected in a synchronous fashion, using registers to equalize delays along each path and by using a simple state machine to throttle the inputs to the clump, based on the actor with the greatest initiation interval. The clumping transformation is akin to the process of converting an asynchronous design, where all actors are connected by FIFOs, into a Globally Asynchronous Locally Synchronous architecture (GALS) [25], where FIFOs connect regions of synchronously connected actors called *clumps*.

2.4.4 I/O Integration

As discussed in the introduction, there is a the tight contract-based integration between DSP Design Module and LabVIEW FPGA, where the actual I/O takes place. Logically, however, we can view I/O operations as being proxied to DSP Design Module via constraints on ports of the actor diagram, that describe the maximum rate at which data is going to be fed from I/O or other FPGA IP modules to the input ports of a DSP diagram, and the minimum rate required at the output ports of the DSP diagram, to feed the I/O or other FPGA IP modules. This separation of concerns allows the LabVIEW FPGA designer to focus on timed behavior, i.e., details of interacting with the I/O, where the effects of time are clearly observable, while the DSP designer can focus on the (untimed) algorithm and let the tools help with the interaction with the timed boundary to LabVIEW FPGA. In the future, we intend to bring the I/O actors directly onto the DSP diagram as special actors that the user can configure, and automatically modify the properties to propagate the constraints onto the surrounding terminals.

2.5 OFDM Transmitter and Receiver Case Study

In this section, we present a case study on the design and implementation of a real-time single antenna OFDM transmitter and receiver using LabVIEW DSP Design Module. The single antenna OFDM link design is based upon the LTE standard [9] with system specifications that include a transmission bandwidth of 5 MHz, 7.68 MSps sampling rate, 512 FFT length, 128 cyclic prefix (CP) length (extended mode), 250 data subcarriers, 50 reference subcarriers, and variable 4/16/64 Quadrature Amplitude Modulation (QAM). The proposed communication system is implemented on an NI PXI Express platform shown in Fig. 2.6, where the transmitter and receiver consist of the following four main components: (a) *PXIe*-8133 Real-time controller

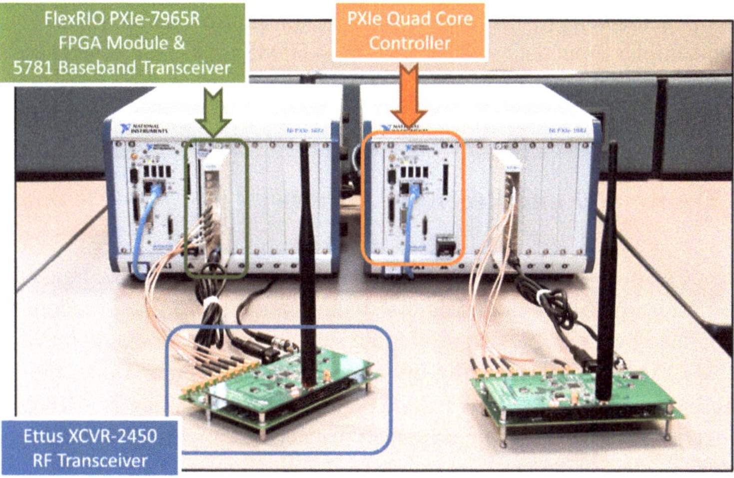

Fig. 2.6 NI PXI Express Real-Time Signal Processing Platform with Ettus Research RF front-end

equipped with a 1.73 GHz quad-core Intel Core i7-820 processor; (b) *PXIe*-7965R FPGA module with a Virtex-5 SX95T FPGA; (c) *NI*-5781 40 MHz baseband transceiver module; and (d) *Ettus Research XCVR*-2450 802.11a/b/g compliant 40 MHz RF transceiver.

Figure 2.7 shows the transmitter and receiver block diagram of the various signal processing tasks. The figure also shows a mapping of the various blocks to the underlying hardware targets and the respective design tools used in their implementation; e.g., the transmitter *Data Bit Generation* block (programmed using LabVIEW Real-time) executes on the PXIe-8133 real-time controller, while the higher rate 512 *IFFT with* 128 *CP Insertion* block (programmed using DSP Design Module) executes on the PXIe-7965R FPGA module. The various data rates associated with the inputs and

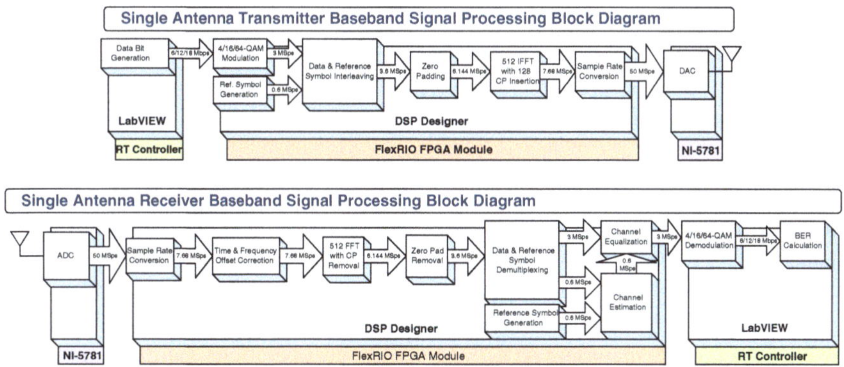

Fig. 2.7 Hardware and software mapping of Transmitter and Receiver block diagrams

outputs of each block are also shown; e.g., the transmitter *Sample Rate Conversion* block up-samples input data streaming at 7.68–50 MSps in order to meet the sample rate constraints of the NI-5781 DAC.

2.5.1 Transmitter and Receiver Overview

We provide an overview of how DSP Design Module implements the transmitter and receiver; refer to [26] for a detailed discussion. In the transmitter, random bytes of data generated by a real-time controller are sent to the FPGA module for Multilevel QAM (M-QAM) [27]. Depending upon the modulation order value, the bytes of data are unpacked into groups of 2, 4, or 6 bits corresponding to 4/16/64-QAM, respectively which are then mapped to their respective complex symbols. After QAM modulation, 250 data symbols are interleaved with 50 reference symbols stored in a look-up table forming an array of 300 interleaved symbols which is then split into two equal groups and padded with zeros forming an array of 512 samples. The 512 samples are passed through a 512 point IFFT block translating the frequency domain samples into the time domain. A 128 point CP is also inserted such that the output of the block consists of 640 samples streaming at 7.68 MSps which is then converted to 50 MSps using two sets of FIR filters. The samples are forwarded to the NI-5781 for digital-to-analog conversion followed by RF up-conversion.

The receiver begins with sample rate down-conversion of the incoming 50 MSps signal (from the ADC) to 7.68 MSps. This is followed by time and carrier frequency offset (CFO) estimation using the blind estimation technique proposed in [28]. After CFO correction, the received OFDM symbol is passed on for CP removal and FFT transformation returning the signal to the frequency domain. Zero pads are then removed, and the reference and data symbols are separated in a de-interleave operation; the data is subsequently processed for channel estimation and channel equalization. Once channel equalization is complete, the data symbol estimates are transferred to the real-time controller at 3 MSps for QAM demodulation and bit error rate calculation.

2.5.2 Hardware Implementation

In addition to the portions of the design implemented in the DSP Design Module, the compilation results include nominal logic implemented in LabVIEW FPGA that manages data transfer across the NI-5781 baseband transceiver and PXIe-7965R FPGA module, and the PXIe-7965R FPGA module and PXIe-8133 controller which is running LabVIEW Real-time. The results also include additional logic to control the NI-5781 module including ADC/DAC read/write operations, sampling frequency configurations, and clock selections.

Table 2.1 shows the summary of the compiled FPGA resource utilization. The first two columns show the various resources available on the PXIe-7965R's Virtex-5 SX95T FPGA and the total number of elements associated with each resource; the percentage utilization by the transmitter and receiver modules are listed in the last two columns. Note that the receiver uses more than twice the registers and LUTs than the transmitter due to significant difference in computational complexity. The DSP diagrams for the transmitter and receiver are configured for 125 MHz clocks, and both successfully met timing during compilation. Figure 2.8 is a screen shot of the OFDM receiver front panel taken during an over-the-air test of the communications link. In addition to carrier frequency, modulation order, and LNA gain controls, a sample 16-QAM signal constellation plot is shown along with two average bit error rate (BER) curves, one taken on a single subframe basis (lower right hand plot), and the other taken over all received subframes (upper right hand plot). The average BER over all subframes converges to an approximate value of 8×10^{-4}, which is viable for a practical implementation.

2.5.3 Design Exploration

Furthermore, the designer can use the analysis framework to explore trade-offs between the throughput and buffer resources used by the implementation. Table 2.2

Table 2.1 Resource utilization of the OFDM transmitter and receiver on a virtex-5 SX95T FPGA

Resource name	Available elements	Transmitter utilization (%)	Receiver utilization (%)
Slices	14720	43.1	79.2
Slice registers	58880	21.6	54.6
Slice LUTs	58880	24.7	57.3
DSP48s	640	2.7	8.3
Block RAM	244	8.2	19.7

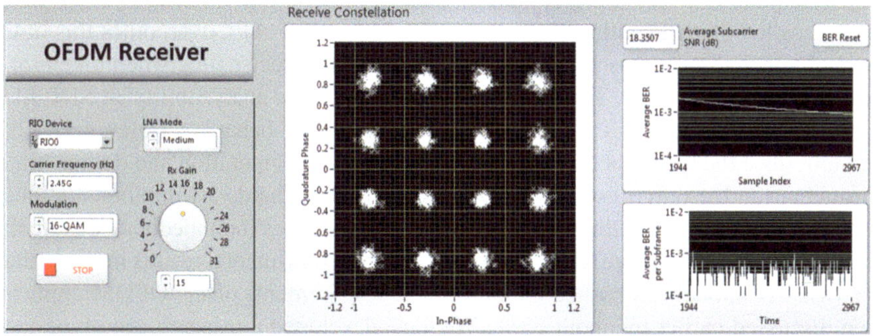

Fig. 2.8 OFDM receiver front panel

Table 2.2 Throughput and buffer size trade-offs for the OFDM transmitter and receiver models

Name	#Actors, #channels	Firings/ iteration	Throughput (MSps)	BlockRAM utilization (%)	Run-time (seconds)
OFDM Transmitter	1114	6250	2	3.6	0.7
			5	4.1	1
			25	6.4	1
			50	8.2	2
OFDM Receiver	4666	25000	15	3.1	40
			50	19.7	39

summarizes these trade-offs for the OFDM transmitter and receiver models. The table reports the number of actors, channels, and firings per iteration for each model. The target throughputs are specified in Mega-Samples-per-second (MSps). The buffer resources required by the design to meet the target throughput are specified in terms of the percentage utilization of the BlockRAM on the PXIe-7965R's Virtex-5 SX95T FPGA target. The CPU run time for the analysis on an Intel Core i7 2.8 GHz processor is also reported.

As expected, lower target throughputs require fewer buffer resources. The graphical environment allows the designer to quickly explore different design configurations and inspect the resource utilization and schedule without generating full hardware implementations. The run times for the more complex receiver models is still less than a minute, which is reasonable for an interactive exploration framework.

2.5.4 Extensions

Following on from these results, we have been able to demonstrate enhancements in resource utilization and performance by using more detailed actor information, in particular data access patterns for SDF actors. The SDF model is limited in its ability to specify how data is accessed in time which often leads to sub-optimal implementations using more resources than necessary. An extension to the SDF model, called Static Dataflow with Access Patterns (SDF-AP), overcomes this limitation by including access patterns which captures the precise time when tokens are read and written at ports. The SDF-AP model retains the analyzability of SDF-like models while accurately capturing the interface timing behavior. References [29, 30] introduce the theory behind the SDF-AP model of computation and demonstrate throughput and resource usage improvements in resulting hardware implementations. [31] presents analysis methods for key execution properties (boundedness, deadlock, throughput, and buffer size) of the SDF-AP model. Experimental results indicate that the SDF-AP model can reduce the buffer size requirements of the OFDM application discussed earlier by about 63 %. We have developed a research prototype that extends DSP Design Module to generate hardware implementations from SDF-AP models. Even though the DSP Design Module application language is expressive enough to capture a variety of communication applications, there are some applications that could benefit from additional modal behavior. To that effect, we are also studying extensions to the application language based on Heterochronous Dataflow (HDF) [32] and Scenario-Aware Data Flow (SADF) [33].

2.6 Summary

In this article, we presented National Instruments LabVIEW DSP Design Module, a framework to specify dataflow models of streaming applications, analyze them, and generate efficient implementations for hardware targets. The Static Dataflow model with extensions for cyclo-static data rates and parameterization is sufficiently expressive in specifying high level streaming applications that underlie today's complex communications systems. The back-end performs key optimizations related to buffer sizing and scheduling. The framework has a rich actor library and facilitates integration of custom-designed IPs from native and third-party sources. Thus, LabVIEW DSP Design Module serves as a system design and exploration framework that enables domain experts, who are not specialized in hardware design, to productively specify applications using high level models of computation and still create realistic deployments, combining efficient hardware synthesis and physical I/O integration. The case study of the OFDM transmitter and receiver shows its viability as a model-based design framework for next generation signal processing and communications systems. Recent advances related to the specification of access patterns and modal behavior in dataflow models are promising extensions to this work.

References

1. Lee, E.A., Messerschmitt, D.G.: Synchronous data flow. Proc. of the IEEE **75**(9), 1235–1245 (1987)
2. Bhattacharyya, S.S., Murthy, P.K., Lee, E.A.: Software Synthesis from Dataflow Graphs. Kluwer Academic Press, Norwell (1996)
3. Eker, J., Janneck, J.W., Lee, E.A., Liu, J., Liu, X., Ludvig, J., Neuendorffer, S., Sachs, S., Xiong, Y.: Taming heterogeneity: the Ptolemy approach. Proc. of the IEEE **91**(1), 127–144 (2003)
4. Lauwereins, R., Engels, M., Adé, M., Peperstraete, J.A.: Grape-II: a system-level prototyping environment for DSP applications. Computer **28**(2), 35–43 (1995)
5. Andrade, H.A., Kovner, S.: Software synthesis from dataflow models for G and LabVIEW. In: Proceedings of the IEEE Asilomar conference on signals, systems, and computers, pp. 1705–1709 (1998)
6. The MathWorks Inc.: Simulink user's guide (2005). http://www.mathworks.com
7. Kee, H., Shen, C.C., Bhattacharyya, S.S., Wong, I., Rao, Y., Kornerup, J.: Mapping parameterized cyclo-static dataflow graphs onto configurable hardware. J. Signal Process. Syst. vol. **66**, pp. 285–301 (2012). doi:10.1007/s11265-011-0599-5

8. Berg, H., Brunelli, C., Lücking, U.: Analyzing models of computation for software defined radio applications. In: International symposium on system-on-chip (SOC), pp. 1–4. Tampere, Finland (2008)
9. 3GPP LTE: The mobile broadband standard (2008) http://www.3gpp.org/
10. Edwards, M., Green, P.: The implementation of synchronous dataflow graphs using reconfigurable hardware. In: Proceedings of FPL '00, pp. 739–748 (2000)
11. Horstmannshoff, J., Meyr, H.: Optimized system synthesis of complex RT level building blocks from multirate dataflow graphs. In: Proceedings of ISSS '99, pp. 38–43 (1999)
12. Jung, H., Lee, K., Ha, S.: Optimized RTL code generation from coarse-grain dataflow specification for fast HW/SW cosynthesis. J. Signal Process. Syst. **52**(1), 13–34 (2008)
13. National Instruments Corp.: LabVIEW FPGA. http://www.ni.com/fpga
14. Xilinx Inc.: System generator for DSP: getting started guide. http://www.xilinx.com
15. Hsu, C.J., Pino, J.L., Hu, F.J.: A mixed-mode vector-based dataflow approach for modeling and simulating lte physical layer. In: Proceedings of the 47th design automation conference, DAC '10, pp. 18–23. ACM, New York, USA (2010)
16. Janneck, J.W.: Open dataflow (OpenDF). http://www.opendf.org/
17. Janneck, J., Miller, I., Parlour, D., Roquier, G., Wipliez, M., Raulet, M.: Synthesizing hardware from dataflow programs: an MPEG-4 simple profile decoder case study. In: IEEE workshop on signal processing systems, pp. 287–292 (2008)
18. Gu, R., Janneck, J.W., Raulet, M., Bhattacharyya, S.S.: Exploiting statically schedulable regions in dataflow programs. In: Proceedings of the 2009 IEEE international conference on acoustics, speech and signal processing, ICASSP '09, pp. 565–568. IEEE Computer Society, Washington, USA (2009)
19. Xilinx Inc.: Xilinx core generator, ISE design suite 12.1. Xilinx Inc. (2010)
20. Stuijk, S., Geilen, M., Basten, T.: Exploring trade-offs in buffer requirements and throughput constraints for synchronous dataflow graphs. In: Proceedings of DAC '06, pp. 899–904 (2006)
21. Bilsen, G., Engels, M., Lauwereins, R., Peperstraete, J.A.: Cyclo-static dataflow. IEEE Trans. Signal Process. **44**(2), 397–408 (1996)
22. Bilsen, G., Engels, M., Lauwereins, R., Peperstraete, J.: Cyclo-static data flow. In: IEEE international conference acoustics, speech, and signal processing. vol. 5, pp. 3255–3258 (1995)
23. Moreira, O.M., Bekooij, M.J.G.: Self-timed scheduling analysis for real-time applications. EURASIP J.n Adv. Signal Process. **2007**(83710), 1–15 (2007)
24. Bhattacharya, B., Bhattacharyya, S.: Parameterized dataflow modeling for DSP systems. IEEE Trans. Signal Process. **49**(10), 2408–2421 (2001)
25. Chapiro, D.M.: Globally-asynchronous locally-synchronous systems. Ph.D. thesis, Stanford University, CA (1984)
26. Andrade, H., Correll, J., Ekbal, A., Ghosal, A., Kim, D., Kornerup, J., Limaye, R., Prasad, A., Ravindran, K., Tran, T.N., Trimborn, M., Wang, G., Wong, I., Yang, G.: From streaming models to FPGA implementations. In: Proceedings of the conference for engineering of reconfigurable systems and algorithms (ERSA-IS). Las Vegas, USA (2012)
27. Proakis, J.: Digital communications, 4th edn. McGraw-Hill Science/Engineering/Math (2000)
28. Sandell, M., van de Beek, J.J., Brjesson, P.O.: Timing and frequency synchronization in OFDM systems using the cyclic prefix. In: Proceedings of international symposium synchronization, pp. 16–19 (1995)
29. Ghosal, A., Limaye, R., Ravindran, K., Tripakis, S., Prasad, A., Wang, G., Tran, T.N., Andrade, H.: Static dataflow with access patterns: semantics and analysis. In: Proceedings of the 49th annual design automation conference, DAC '12, pp. 656–663. ACM, New York, USA (2012)
30. Tripakis, S., Andrade, H., Ghosal, A., Limaye, R., Ravindran, K., Wang, G., Yang, G., Kornerup, J., Wong, I.: Correct and non-defensive glue design using abstract models. In: Proceedings of the seventh IEEE/ACM/IFIP international conference on hardware/software codesign and system synthesis, CODES+ISSS '11, pp. 59–68. ACM, New York, USA (2011)
31. Ravindran, K., Ghosal, A., Limaye, R., Wang, G., Yang, G., Andrade, H.: Analysis techniques for static dataflow models with access patterns. In: Proceedings of the 2012 conference on design and architectures for signal and image processing, DASIP '12 (2012)

32. Girault, A., Lee, B., Lee, E.A.: Hierarchical finite state machines with multiple concurrency models. IEEE Trans. Comput. Aided Des. **18**(6), 742–760 (1999)
33. Theelen, B.D., Geilen, M.C.W., Basten, T., Voeten, J.P.M., Gheorghita, S.V., Stuijk, S.: A scenario-aware data flow model for combined long-run average and worst-case performance analysis. In: Proceedings of MEMOCODE'06, pp. 185–194 (2006)

Chapter 3
Dataflow-Based, Cross-Platform Design Flow for DSP Applications

Zheng Zhou, Chung-Ching Shen, William Plishker and Shuvra S. Bhattacharyya

Abstract Dataflow methods have been widely explored over the years in the digital signal processing (DSP) domain to model, design, analyze, implement, and optimize DSP applications, such as applications in the areas of audio and video data stream processing, digital communications, and image processing. DSP-oriented dataflow methods provide formal techniques that facilitate software design, simulation, analysis, verification, instrumentation and optimization for exploring effective implementations on diverse target platforms. As the landscape of embedded platforms becomes increasingly diverse, a wide variety of different kinds of devices, including graphics processing units (GPUs), multicore programmable digital signal processors (PDSPs), and field programmable gate arrays (FPGAs), must be considered to thoroughly address the design space for a given application. In this chapter, we discuss design methodologies, based on the core functional dataflow (CFDF) model of computation, that help engineers to efficiently explore such diverse design spaces. In particular, we discuss a CFDF-based design flow and associated design methodology for efficient simulation and implementation of DSP applications. The design flow supports system formulation, simulation, validation, cross-platform software implementation, instrumentation, and system integration capabilities to derive optimized signal processing implementations on a variety of platforms. We provide a comprehensive specification of the design flow using the lightweight dataflow (LWDF) and targeted dataflow interchange format (TDIF) tools, and demonstrate it

Z. Zhou (✉) · C.-C. Shen · W. Plishker · S. S. Bhattacharyya
Department of Electrical and Computer Engineering and Institute for Advanced Computer Studies, University of Maryland, College Park, Maryland, USA
e-mail: zhengzho@umd.edu

C.-C. Shen
e-mail: ccshen@umd.edu

W. Plishker
e-mail: plishker@umd.edu

S. S. Bhattacharyya
e-mail: ssb@umd.edu

A. Sangiovanni-Vincentelli et al. (eds.), *Embedded Systems Development*,
Embedded Systems 20, DOI: 10.1007/978-1-4614-3879-3_3,

with case studies on CPU/GPU and multicore PDSP designs that are geared towards fast simulation, quick transition from simulation to the implementation, high performance implementation, and power-efficient acceleration, respectively.

3.1 Introduction

As embedded processing platforms become increasingly diverse, designers must evaluate trade-offs among different kinds of devices such as CPUs, graphics processing units (GPUs), multicore programmable digital signal processors (PDSPs), and field programmable gate arrays (FPGAs). The diversity of relevant platforms is compounded by the trend towards integrating different kinds of processors onto heterogeneous multicore devices for DSP (e.g., see [1]). Such heterogeneous platforms help designers to simultaneously achieve manageable cost, high power efficiency, and high performance for critical operations. However, there is a large gap from the simulation phase to the final implementation. Simulation is used extensively in the early stage of system design for high-level exploration of design spaces and fast validation. In contrast, in the implementation phase, there is strong emphasis platform dependent issues, performance centric optimization, and tuning low-level implementation trade-offs. A seamless design flow is needed to help designers to effectively bridge this gap.

Dataflow models of computation have been widely used in the design and implementation of DSP applications, such as applications for audio and video stream processing, digital communications, and image processing (e.g., see [1, 2]). These applications often require real-time processing capabilities and have critical performance constraints. Dataflow provides a formal mechanism for specifying DSP applications, imposes minimal data-dependency constraints in specifications, and is effective in exposing and exploiting task or data level parallelism for achieving high performance implementations.

A *dataflow graph* is a directed graph, where vertices (*actors*) represent computational functions (*actors*), and edges represent first-in-first-out (FIFO) channels for storing data values (*tokens*) and imposing data dependencies between actors. In DSP-oriented dataflow models of computation, actors can typically represent computations of arbitrary complexity as long as the interfaces of the computations conform to dataflow semantics. That is, dataflow actors produce and consume data from their input and output edges, respectively, and each actor executes as a sequence of discrete units of computation, called *firings*, where each firing depends on some well-defined amount of data from the input edges, and produces some well-defined amount of data onto the output edges of the associated actor [3].

Dataflow modeling has several advantages, which facilitate its use as a single, unified model across the simulation phase and implementation phases of the design flow. First, dataflow utilizes simple, standard, and loosely coupled interfaces (in the form of FIFOs), which can be easily implemented on individual platforms and across different platforms. Second, the structure of actor interfaces can be cleanly separated from the internal operation of actors. As long as the predefined amounts of data from the input edges are consumed and the predefined amounts of data to the output edges are produced on each firing, the implementation of an actor can be changed without violating key higher level properties of the enclosing dataflow model. Third, dataflow models enable a wide variety of powerful techniques for design analysis and optimization—e.g., for evaluating or optimizing performance, memory usage, throughput and latency [1].

Based on *Core Functional Dataflow* (*CFDF*) [4] , two complementary tools have been developed in recent years. First, the *lightweight dataflow* (*LWDF*) programming methodology [5] provides a "minimalistic" approach for integrating coarse grain dataflow programming structures into DSP simulation for fast system formulation, validation and profiling with arbitrary languages. Second, the *targeted dataflow interchange format* (*TDIF*) framework [6] provides cross-platform actor design support, and the integration of (1) code generation for programming interfaces, and (2) low level customization for implementations targeted to homogeneous and heterogeneous platforms alike. In this chapter, we present a novel dataflow-based design flow that builds upon on both LWDF and TDIF to allow rapid transition from simulation to optimized implementations on diverse platforms. In this chapter, we present this design flow, and provide case studies to demonstrate its application.

3.2 Background

In this section, we review background that is needed to introduce our design flow, which is introduced in the next section.

3.2.1 CFDF Dataflow Model

Enable-Invoke Dataflow (EIDF) is a dataflow model [4] that facilitates the design of applications with structured dynamic behavior. EIDF divides actors into sets of *modes*. Each mode, when executed, consumes and produces a fixed number of tokens. The fixed behavior of a mode provides structure that can be exploited by analysis and optimization tools, while dynamic behavior can be achieved by switching between modes at run-time.

Each mode is defined by an *enable method* and an *invoke method*, which correspond, respectively, to testing for sufficient input data and executing a single quantum ("invocation") of execution for a given actor. After a successful invocation, the invoke method returns a set of admissible *next modes*, any of which may be then checked for readiness using the enable method and then invoked, and so on. By returning a *set* of possible next modes (as opposed to being restricted to a single next mode), a designer can model non-deterministic applications in which execution can proceed down multiple alternative paths.

In the implementation of dataflow tools, functionality corresponding to the enable and invoke methods is often interleaved—for example, an actor firing may have computations that are interleaved with blocking reads of data that provide successive inputs to those computations. In contrast, there is a clean separation of enable and invoke capabilities in EIDF. This separation helps to improve the predictability of an actor invocation (since availability of the required data can be guaranteed in advance by the enable method), and in prototyping efficient scheduling and synthesis techniques (since enable and invoke functionality can be called separately by the scheduler).

CFDF is a special case of the EIDF model. Recall that in EIDF, the invoking function in general returns a set of valid next modes in which the actor can subsequently be invoked. This allows for non-determinism as an actor can be invoked in any of the valid modes within the next-mode set. In the deterministic CFDF model, actors must proceed deterministically to one particular mode of execution whenever they are enabled. Hence, the invoke function should return only a single valid mode of execution instead of a set of arbitrary cardinality. In other words, CFDF is the model of computation that results when EIDF is restricted so that the set of next modes always has exactly one element. With this restricted form of invoke method, only one mode can meaningfully be interrogated by the enable method, ensuring that the application is deterministic.

3.2.2 Lightweight Dataflow

Lightweight dataflow (*LWDF*) [5] is a programming approach that allows designers to integrate various dataflow modeling approaches relatively quickly and flexibly into existing design methodologies and processes. LWDF is designed to be minimally intrusive on existing design processes and requires minimal dependence on specialized tools or libraries. LWDF can be combined with a wide variety of dataflow models to yield a lightweight programming method for those models. In our design flow, the LWDF is used for system simulation.

In LWDF, an actor has an *operational context*, which encapsulates the following entities related to an actor design [7]:

- Actor parameters.
- Local actor variables—variables whose values store temporary data that does not persist across actor firings.
- Actor state variables—variables whose values do persist across firings.
- References to the FIFOs corresponding to the input and output ports (edge connections) of the actor as a component of the enclosing dataflow graph.
- Terminal modes: a (possibly empty) subset of actor modes in which the actor cannot be fired.

In LWDF, the operational context for an actor also contains a *mode variable* whose value stores the next CFDF mode of the actor and persists across firings. The

LWDF operational context also includes references to the invoke function and enable function of the actor. The concept of terminal modes, defined above, can be used to model finite subsequences of execution that are "re-started" only through external control (e.g., by an enclosing scheduler). This is implemented in LWDF by extending the standard CFDF enable functionality such that it unconditionally returns `false` whenever the actor is in a terminal mode.

3.2.3 The Targeted DIF Framework

TDIF [6] is a two layer framework with each design involving a system layer and actor layer. This two layer architecture helps to separate system design from actor design, which reduces total design complexity and enhances modularity. Based on this two layer approach, TDIF provides a systematic solution for developing DSP applications on diverse platforms. The CFDF dataflow model used in TDIF provides capabilities for formal analysis of system behavior, while providing the programmer with design flexibility and multi-level implementation tuning.

An illustration of the TDIF environment and its associated design flow is shown in Fig. 3.1. By applying this methodology, the designer can focus on design, implementation and optimization for dataflow actors and experiment with alternative scheduling strategies for the targeted platforms based on programming interfaces that are automatically generated from the TDIF tool. These automatically-generated interfaces provide structured design templates for the designer to follow in order to generate dataflow-based actors that are formally integrated into the overall synthesis tool.

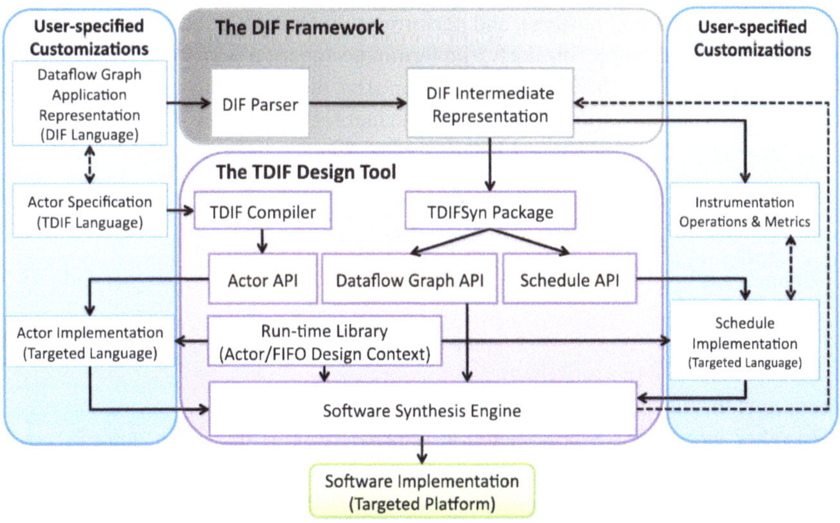

Fig. 3.1 TDIF-based design and synthesis flow

In Fig. 3.1, dashed lines indicate design considerations that need to be taken into account jointly to achieve maximum benefit from TDIF-based system design. In our proposed design flow, the TDIF framework is used primarily in the implementation phase (not for simulation).

The TDIF tool is based on four software packages—the *TDIF compiler*, *TDIFSyn* software synthesis package, *TDIF run-time library*, and *Software Synthesis Engine*. Interactions among these packages are illustrated in Fig. 3.1.

3.3 From Simulation to Implementation

In this section, we elaborate on the process of applying our proposed design flow based on LWDF and TDIF, and we discuss useful features of the design flow.

3.3.1 Step 1: System Formulation

In the first step of the design flow, the targeted DSP application is modeled in terms of CFDF semantics. Decisions are made on how to decompose the application specification into actors, along with the functionality of and parameterization associated with each actor.

Next, we design FIFOs and actors or select such design components from predefined libraries. Actor design in LWDF includes four interface functions—the *construct*, *enable*, *invoke*, and *terminate* functions. The construct function can be viewed as a form of object-oriented constructor, which connects an actor to its input and output edges (FIFO buffers), and performs any other pre-execution initialization associated with the actor. Similarly, the terminate function performs any operations that are required for "closing out" the actor after the enclosing graph has finished executing (e.g., deallocation of actor-specific memory or closing of associated files).

In the enable function for an LWDF actor a, a `true` value is returned if

$$\mathit{population}(e) \geq \mathit{cons}(a, m, e) \text{ for all } e \in \mathit{inputs}(a); \tag{3.1}$$

$$\mathit{population}(e) \leq (\mathit{capacity}(e) - \mathit{prod}(a, m, e)) \text{ for all } e \in \mathit{outputs}(a); \text{ and} \tag{3.2}$$

$$m \notin \tau(a), \tag{3.3}$$

where m is the current mode of a.

In other words, the enable function returns `true` if the given actor is not in a terminal mode, and has sufficient input data to execute the current mode, and the output edges of the actor have sufficient data to store the new tokens that are produced when the mode executes. An actor can be invoked at a given point of time if the enable function is `true`-valued at that time.

In the invoke function of an LWDF actor a, the operational sequence associated with a single invocation of a is implemented. Based on CFDF semantics, an actor proceeds deterministically to one particular mode of execution whenever it is enabled, and in any given mode, the invoke method of an actor should consume data from at least one input or produce data on at least one output (or both) [4]. Note that in case an actor includes state, then the state can be modeled as a self-loop edge (a dataflow edge whose source and sink actors are identical) with appropriate delay, and one or more modes can be defined that produce or consume data only from the self-loop edge. Thus, modes that affect only the actor state (and not the "true" inputs or outputs of the actor) do not fundamentally violate CFDF semantics, and are therefore permissible in LWDF.

After completing actor design and selection, we focus an analogous process for FIFOs. FIFO design for LWDF dataflow edge implementation is orthogonal to the design of dataflow actors in LWDF. That is, by using LWDF, application designers can focus on design of actors and mapping of edges to lower level communication protocols through separate design processes (if desired) and integrate them later through well-defined interfaces. Such design flow separation is useful due to the orthogonal objectives, which center around computation and communication, respectively, associated with actor and FIFO implementation.

Standard FIFO operations in LWDF include operations that perform the following tasks:

- Create a new FIFO with a specified capacity;
- Read and write tokens from and to a FIFO;
- Check the capacity of a FIFO;
- Check the number of tokens that are currently in a FIFO;
- Deallocate the storage associated with a FIFO (e.g., for dynamically adapting graph topologies or, more commonly, as a termination phase of overall application execution).

The buffer sizes used during simulation and implementation are determined based on the scheduling strategies that are employed. Our proposed design flow facilitates experimentation with alternative scheduling strategies to help designers explore trade-offs involving buffer sizes and other relevant implementation metrics, such as latency and throughput. More details on scheduling are provided later in this section.

3.3.2 Step 2: System Validation and Profiling

After system formulation, we need a schedule to validate the correctness of the system and the behavior of each actor. A CFDF *canonical schedule* [8] can be employed for this purpose. A canonical schedule repeatedly traverses the list of actors in a CFDF graph according to some pre-defined sequence. When an actor is visited during this traversal process, its enable function is first called. If the enable function returns `true`, the invoke function for the actor is subsequently called; otherwise, firing of the actor is effectively "skipped", and traversal continues with the next actor in the sequence.

During simulation, profiling information for each actor is obtained to extract relevant performance characterizations that will be employed to inform the implementation phase. For example, execution time statistics for each actor mode are extracted. Such profiling can help to identify modes that are bottlenecks of individual actors, and actors that are bottlenecks of the overall system.

3.3.3 Step 3: System Optimization

In this step, we enter the implementation phase, which is where TDIF comes into play. There are two main kinds of optimization techniques supported in the TDIF framework. One is *cross-platform implementation* for actor-level optimization, and the other is *scheduling and mapping* for system or subsystem optimization.

After we identify the bottleneck actors, cross-platform implementation of actors allows designers to efficiently experiment with alternative actor realizations on different kinds of platforms, such as GPUs, multicore PDSPs, and FPGAs, to help derive a platform or mix of platforms that will be strategic in terms of the given design constraints (e.g., constraints involving cost, performance, and energy consumption). During this process, much of the code from the simulation phase can be reused. Only the functionality associated with selected actor modes (e.g., bottleneck modes of bottleneck actors) needs to be rewritten or selected from available platform-specific libraries.

The TDIF environment currently supports C-like programming languages—i.e., languages that are targeted to CPU, GPU and multicore PDSP platforms. The GPU-based capabilities of TDIF are currently oriented towards NVIDIA GPUs, based on the CUDA programming language [9], which can be viewed as an extension of C. The multicore PDSP capabilities currently in TDIF are oriented towards Texas Instruments (TI) PDSP devices, and are interoperable with the multithreading libraries provided by TI [10].

TDIF also provides a library of FIFO implementations that are optimized for different platforms. These FIFOs all adhere to standard operations defined in LWDF so that they can be integrated in a manner that is consistent with the CFDF graph model from the simulation phase. After simulation-mode FIFOs are mapped into platform-specific FIFOs, optimized actors can communicate in a manner that is efficient, and consistent with the designer's simulation model.

The scheduling strategy employed determines the execution order of the actors while the mapping process, which is typically coupled closely with scheduling, determines which resource each actor is executed on. TDIF provides the *generalized schedule tree* (*GST*) [11] representation to facilitate implementation of and experimentation with alternative scheduling and mapping schemes for system optimization. GSTs are ordered trees with leaf nodes and internal nodes. An internal node of a GST in TDIF represents iteration control (e.g., a loop count) for an iteration construct that is to be applied when executing the associated subtree. On the other hand, a GST leaf node includes two pieces of information that are used to carry out individual actor firings—one is an actor of the associated dataflow graph, and the other is mapping information associated with the actor. The GST representation provides designers with a common interface through with topological information and algorithms for ordered trees can be applied to access and manipulate schedule elements.

Execution of a GST involves traversing the tree to iteratively enable (and then execute, if appropriate) actors that correspond to the schedule tree leaf nodes. Note that if actors are not enabled, the GST traversal simply skips their invocation. Subsequent schedule rounds (and thus subsequent traversals of the schedule tree) will generally revisit actors that were unable to execute in the current round.

For schedule construction in the implementation phase, the *CFDF graph decomposition* approach of [12] is integrated in the TDIF framework. This approach allows designers to decompose a CFDF graph into a set of SDF subgraphs. Each SDF subgraph can be scheduled by existing static scheduling algorithms, such as an APGAN-based scheduler [13]. The GST schedule trees that result from scheduling these SDF subgraphs are then systematically combined into a single, "execution-rate-balanced" GST using profiling and instrumentation techniques that are discussed in next section.

3.3.4 Step 4: System Verification and Instrumentation

Given a GST together with a set of actor and FIFO implementations, the TDIF framework automatically generates a complete CFDF graph implementation for the target platform. Performance metrics, such as latency and throughput, can then be evaluated using platform-specific simulation tools, or using execution on actual target hardware.

The TDIF framework also provides an instrumentation approach to help assess generated implementations. Such instrumentation facilitates experimentation with and tuning of alternative scheduling and mapping techniques. The instrumentation approach integrated in TDIF also facilitates trade-off assessment across different implementation metrics, and helps designers steer implementations towards effective solutions.

Our approach to instrumentation in TDIF is designed to support the following features: (1) no change in functionality (instrumentation directives should not change application functionality); (2) operations for adding and removing instrumentation points should be performed by designers in a way that is external to actors (i.e., does not interfere with or require modification of actor code); and (3) instrumentation operations should be *modular* so that they can be mixed, matched, and migrated with ease and flexibility.

In *schedule tuning mode*, TDIF allows designers to augment the GST representation with functional modules, encapsulated as *instrumentation nodes*, which are dedicated to instrumentation tasks. Like iteration nodes, instrumentation nodes are incorporated as internal nodes. We refer to GSTs that are augmented with instrumentation nodes as *instrumented GSTs* (*IGSTs*). The instrumentation tasks associated with an instrumentation node are in general applied to the corresponding IGST sub-tree.

An IGST allows software synthesis for a schedule together with instrumentation functionality that is integrated in a precise and flexible format throughout the schedule. Upon execution, software that is synthesized from an IGST produces profiling data (e.g., related to memory usage, performance or power consumption) along with the output data that is generated by the source application. Modeling techniques, metrics, and measurements related to the coding efficiency of TDIF-based implementations are discussed in [14].

3.3.5 Determining Buffer Sizes

During simulation, estimated buffer size bounds are provided by the designer. If the these bounds are not sufficient to keep the graph running without deadlock, then simulation is terminated with an appropriate diagnostic message. The designer can then increase selected buffer sizes, and retry the simulation. This process is repeated until the simulation completes for the desired number of iterations, or the buffer size constraints of the design are exhausted. In the latter case, the designer needs to re-examine the system for sample-rate inconsistencies (unbounded buffer sizes), and re-design the system to reduce buffer sizes. Such an iterative process is needed for general CFDF specifications, which are highly expressive, and therefore do not provide the kinds of guaranteed buffer size bounds that are available with less expressive models, such as synchronous dataflow (SDF) or cyclo-static dataflow (CSDF) [2, 15].

During the implementation phase, subgraphs corresponding to specialized models, such as SDF or CSDF, can be extracted from the CFDF specification and optimized with appropriate guaranteed-buffer-bound algorithms. In particular, we apply the APGAN scheduling technique to optimize buffer sizes for SDF subgraphs [13]. Furthermore, quasi-static scheduling techniques, such as the CFDF *mode grouping* technique, can be applied to provide buffer optimization for CFDF specifications that contain certain kinds of dynamic dataflow structures (e.g., see [16]).

3.3.6 Discussion

Integration of the following four features distinguish the CFDF-based design flow developed in this chapter:

1. Support for dynamic dataflow behavior in the system;
2. Use of a common formal model for simulation and implementation, which provides rapid transition between the simulation and implementation phases, and promotes consistency between simulation and implementation versions of a design;

3. Support for diverse target platforms;
4. Support for significant code reuse across the simulation and implementation phases.

One limitation of TDIF is that due to the high expressive power of the underlying CFDF model of computation, key analysis and verification problems are undecidable for general TDIF-based applications (e.g., see [8]). However, the CFDF model helps to expose and exploit subsystems within an overall dataflow graph specification that adhere to more specialized (static) dataflow modeling techniques, so that decidable verification properties can be exploited at a "local" level on those subsystems. Some work has also been developed on quasi-static scheduling of CFDF specifications, where fully-verified schedules for static dataflow subsystems are integrated systematically with global dynamic schedulers [16]. Efficient and reliable quasi-static and dynamic scheduling of CFDF specifications are useful directions for further investigation.

3.4 Case Study 1:CPU/GPU

To demonstrate our design flow, we first experiment with an image processing application centered on Gaussian filtering. Two-dimensional Gaussian filtering is a common kernel in image processing that is used for preprocessing. Gaussian filtering can be used to denoise an image or to prepare for multiresolution processing. A Gaussian filter is a filter whose impulse response is a Gaussian curve, which in two dimensions resembles a bell.

For filtering in digital systems, the continuous Gaussian filter is sampled in a window and stored as coefficients in a matrix. The filter is convolved with the input image by centering the matrix on each pixel, multiplying the value of each entry in the matrix with the appropriate pixel, and then summing the results to produce the value of the new pixel. This operation is repeated until the entire output image has been created.

The size of the matrix and the width of the filter may be customized according to the application. A wide filter will remove noise more aggressively but will smoothen sharp features. A narrow filter will have less of an impact on the quality of the image, but will be correspondingly less effective against noise.

It should also be noted that the tiles indicated in Fig. 3.2 do vary somewhat between edges. Gaussian filtering applied to tiles must consider a limited neighborhood around each tile (called a *halo*) for correct results. Therefore, tiles produced by `bmp_file_reader` overlap, while the halo is discarded after Gaussian filtering. As a result, non-overlapping tiles form the input to `bmp_file_writer`.

Figure 3.2 shows a simple application based on Gaussian filtering. It reads bitmap files in tile chunks, inverts the values of the pixels of each tile, runs Gaussian filtering on each inverted tile, and then writes the results to an output bitmap file. The main processing pipeline is single-rate in terms of tiles, and can be statically scheduled, but

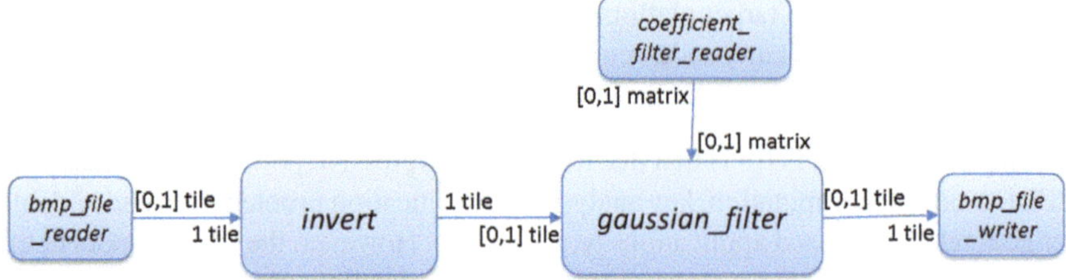

Fig. 3.2 Dataflow graph of an image processing application for Gaussian filtering

after initialization and end-of-file behavior is modeled, there is conditional dataflow behavior in the application graph, which is represented by square brackets in the figure.

Such conditional behavior arises, first, because the Gaussian filter coefficients are programmable to allow for different standard deviations. The coefficients are set once per image—`coefficient_filter_reader` produces a coefficient matrix for only the first firing. To correspond to this behavior, the `gaussian_filter` actor consumes the coefficient matrix only once, and each subsequent firing processes tiles. Such conditional firing also applies to `bmp_file_reader`, which produces tiles until the end of the associated file is reached.

As shown in Fig. 3.2, our dataflow graph of the image processing application for Gaussian filtering is specified as a CFDF graph. The graph includes five actors: `bmp_file_reader`, `coefficient_filter_reader`, `invert`, `gaussian_filter`, and `bmp_file_writer`. We first use the LWDF programming methodology, integrated with the C language, to construct the system for simulation.

3.4.1 Simulation

In our design, the `bmp_file_reader` actor is specified using two CFDF modes, and one output FIFO. The two modes are the `process` mode and the `inactive` mode. It is the actor programmer's responsibility to implement the functionality of each mode. In the `process` mode, the `bmp_file_reader` reads image pixels of a given tile and the corresponding header information from a given bitmap file, and produces them to its output FIFOs. Then the actor returns the `process` mode as the mode for its next firing. This continues for each firing until all of the data has been read from the given bitmap file. After that, the actor returns the `inactive` mode, which is a *terminal mode*. Arrival at a terminal mode indicates that the actor cannot be fired anymore until its current mode is first reset externally (e.g., by the enclosing scheduler).

The `coefficient_filter_reader` actor is also specified in terms of two modes and one output FIFO. The two modes are again labeled as the `process` mode and the `inactive` mode, and again, the `inactive` mode is a terminal mode. On each firing when it is not in the `inactive` mode, the `coefficient_filter_reader` actor reads filter coefficients from a given file, stores them into a *filter coefficient vector* (*FCV*) array, and produces the coefficients onto its output FIFO. The FCV V has the form

$$V = (\texttt{sizeX}, \texttt{sizeY}, c_0, c_1, \ldots, c_{n-1}), \tag{3.4}$$

where `sizeX` and `sizeY` denote the size of the FCV represented in two dimensional format; each c_i represents a coefficient value; and $n = \texttt{sizeX} \times \texttt{sizeY}$. After firing, the actor returns the `process` mode if there is data remaining in the input file; otherwise, the actor returns the `inactive` mode.

The `bmp_file_writer` actor contains only a single mode and one input FIFO. The single mode is called the `process` mode. Thus, the actor behaves as an SDF actor. On each firing, the `bmp_file_writer` actor reads the processed image pixels of the given tile and the corresponding header information from its input FIFOs, and writes them to a bitmap file, which can later be used to display the processed results. The actor returns the `process` mode as the next mode for firing.

The `gaussian_filter` actor contains one input FIFO, one output FIFO and two modes: the store coefficients (`STC`) mode and the `process` mode. On each firing in the `STC` mode, the `gaussian_filter` actor consumes filter coefficients from its coefficient input FIFO, caches them inside the actor for further reference, and then returns the `process` mode as the next mode for firing. In the `process` mode, image pixels of a single tile will be consumed from the tile input FIFO of the actor, and the cached filter coefficients will be applied to these pixels. The results will be produced onto the tile output FIFO. The actor then returns the `process` mode as the next mode for firing. To activate a new set of coefficients, the actor must first be reset, through external control, back to the `STC` mode.

The `invert` actor also contains a single mode called the `process` mode, and contains one input FIFO and one output FIFO. Because it has only one mode, it can also be viewed as an SDF actor. On each firing, the `invert` actor reads the image pixels of the given tile from its input FIFOs, inverts the color of the image pixels, and writes the processed result to its output FIFO. The actor always returns the `process` mode as the next mode for firing.

After designing the actors, as described above, we connect the actors with the appropriate FIFOs. For our simulation setup, we use 256×256 images decomposed into 128×128 tiles, and filtered with different sizes of matrices for Gaussian filter coefficients. The canonical scheduler (see Sect. 3.3.2) is used to run the simulation on 3GHz Intel Xeon processors. The profiling results are reported in Table 3.1. As can be observed from this table, increases in the matrix size lead to increases in the processing time for the Gaussian filter and the overall application. Furthermore, the Gaussian filter actor accounts for most of the processing time in the application in all cases. Thus, if we can optimize the Gaussian filter actor, the performance of the overall application will be enhanced.

Table 3.1 Execution time of the `gaussian_filter` (GF) actor and the Gaussian filtering application (App) during simulation

Filter size	5 × 5	11 × 11	21 × 21	25 × 25	37 × 37
GF. SIM (ms)	50	280	1080	1540	3310
App. SIM (ms)	70	295	1100	1550	3340
Percentage	**71.4 %**	**95 %**	**98.2 %**	**99.3 %**	**99.1 %**

3.4.2 Implementation

From the experiments discussed in the previous section, we identified the bottleneck actor to be the Gaussian filtering actor. To improve the performance of this actor, we apply the cross-platform implementation features of TDIF. In particular, we use TDIF to experiment with a new version of the implementation in which the Gaussian filtering actor is executed on a graphics processing unit (GPU).

GPUs provide a class of high performance computing platforms that provide high peak throughput processing for certain kinds of regularly structured computations [17]. Typically, a GPU architecture is structured as an array of hierarchically connected cores as shown in Fig. 3.3. Cores tend to be lightweight as the GPU will instantiate many of them to support massively parallel graphics computations. Some of the memories are small and scoped for access to small numbers of cores, but can be read or written in one or just a few cycles. Other memories are larger and accessible by more cores, but at the cost of longer read and write latencies.

Using TDIF, we explore the use of GPUs to accelerate the `gaussian_filter` actor. We employ an NVIDIA GTX 285 GPU and employ the CUDA programming environment to specify the internal functionality of the `gaussian_filter` actor for GPU acceleration. This CUDA based actor implementation is integrated systematically into the overall application-level CFDF graph through the TDIF design environment. We apply actor-level vectorization to exploit data parallelism within the actor on the targeted GPU.

Fundamentals of vectorized execution for dataflow actors have been developed by Ritz [18], and explored further by Zivojnovic [19], Lalgudi [20], and Ko [21]. In such vectorization, multiple firings of the same actor are grouped together for execution to reduce the rate of context switching, enhance locality, and improve processor pipeline utilization. On GPUs, groups of vectorized firings can be executed concurrently to achieve parallel processing across different invocations of the associated actor. Each instance of a "vectorized actor" may be mapped to an individual thread or process, allowing the replicated instances to be executed in parallel.

An application developer may consider vectorization within and across actors while writing kernels for CUDA acceleration. In TDIF, the actor interface need not change as the vectorization degree changes, which makes it relatively easy for designers to start with the programming framework provided by CUDA and wrap the resulting vectorized kernel designs in individual modes of an actor for integration at the dataflow graph level.

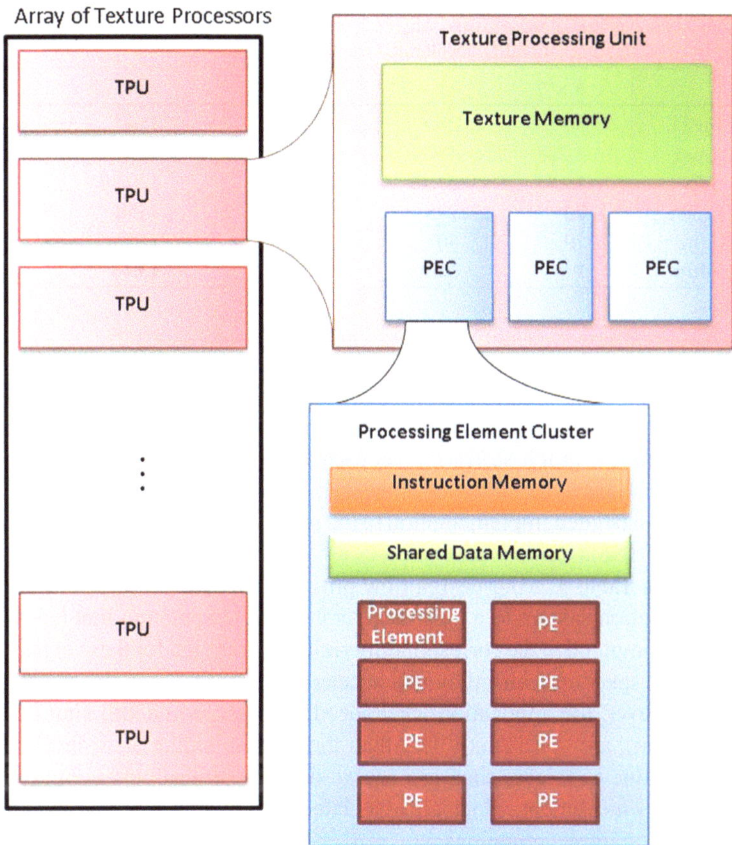

Fig. 3.3 A typical GPU architecture

In the GPU-targeted version of our Gaussian filtering application, a CUDA kernel is developed to accelerate the core Gaussian filtering computation (the `process` mode), and each thread is assigned to a single pixel, which leads to a set of parallel independent tasks. The threads are assembled into blocks to maximize data reuse. Each thread uses the same matrix for application to the local neighborhood, and there is significant overlap in the neighborhoods of the nearby pixels. To this end, the threads are grouped by tiles in the image. Once the kernel is launched, threads in a block cooperate to load the matrix, the tile to be processed, and a surrounding neighborhood of points. The image load itself is vectorized to ensure efficient bursting from memory. Because CUDA recognizes the contiguous accesses across threads, the subsequent image processing operations induce vectorized accesses to global memory.

We use the same canonical scheduler in the GPU implementation that we used in in the simulation phase. The performance of our Gaussian filtering application in simulation and GPU-accelerated implementation is compared to demonstrate the

Table 3.2 Execution time of the `gaussian_filter` (GF) actor and the Gaussian filtering application (App) in simulation and GPU-accelerated implementation

Filter size	5 × 5	11 × 11	21 × 21	25 × 25	37 × 37
GF. SIM (ms)	50	280	1080	1540	3310
GF. GPU (ms)	4.228	4.874	10.257	12.759	21.72
GF. **Speedup**	**11.83**	**57.45**	**105.29**	**120.70**	**152.39**
App. SIM (ms)	70	295	1100	1550	3340
App. GPU (ms)	70	80	140	115	130
App. **Speedup**	**1**	**3.69**	**7.86**	**13.48**	**25.69**

ability of our design flow to support cross-platform actor implementation exploration in a manner that is systematically coupled with the simulation-level application model. We use the same experimental setup—in terms of input and output images and overall dataflow graph structure—as used in the simulation. To accelerate the Gaussian Filtering actor, we applied an NVIDIA GTX 285 running CUDA 3.1 and compared the associated implementation to the simulation system. The measurement results are reported in Table 3.2.

As shown in Table 3.2, our design flow provides flexible and efficient transition from the simulation system to a GPU-accelerated implementation that has superior performance compared to the corresponding simulation design for these experiments. The actor-level speedup realized by this acceleration process is in the range of $10X$ to $100X$. However, the application-level speedup levels, while still significant (up to $25X$ speedup), are consistently less than the corresponding actor-level speedup levels. This is due to factors such as context switch overhead and communication cost for memory movement, which are associated with overall schedule coordination in the application implementations.

3.5 Case Study 2:Multicore PDSP

A programmable digital signal processor (PDSP) is a specialized kind of microprocessor with an architecture optimized for the operational needs of real-time digital signal processing [22]. In multicore PDSPs, such as those available in the Texas Instruments TMS320C6678 family of fixed point and floating point processors [10], multiple PDSP cores are integrated on a single chip and connected with shared memory. A typical multicore PDSP is shown in Fig. 3.4, where each core has private L1 cache as well as L2 cache. All cores share on-chip SRAM and DRAM. High peak throughput can be achieved if all PDSP cores can operate parallel.

Multicore PDSPs are increasingly employed in wireless communication systems, such as systems for software defined radio (SDR). Figure 3.5 shows a CFDF graph model for the mp-sched benchmark [23], which is representative of an important class of digital processing subsystems for wireless communication, and is designed for use with the GNU Radio environment [24]. There are two paths between *SRC* and

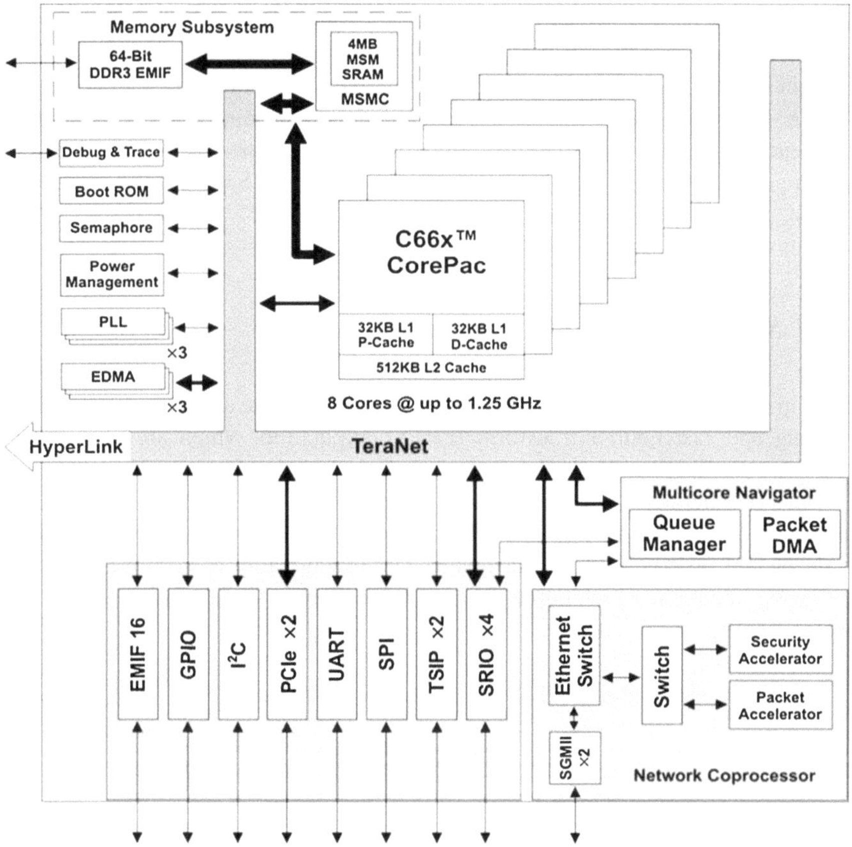

Fig. 3.4 Block diagram of TI TMS320C6678 8-core PDSP device

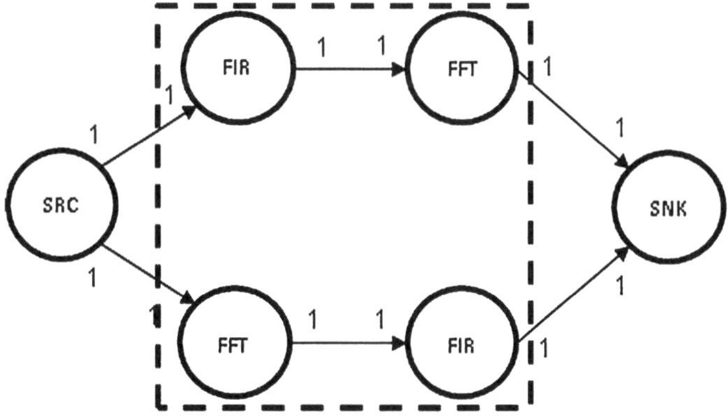

Fig. 3.5 An illustration of the mp-sched benchmark

SNK, which represent two different signal processing procedures on the incoming signals. In the upper path, from *SRC* to *SNK*, the signal is first filtered in the time-domain and then transformed to the frequency-domain. In the lower path, from *SRC* to *SNK*, the signal is first transformed to the frequency-domain and then filtered in the frequency-domain. We illustrate the utility of our design flow to conduct simulation and implementation of the mp-sched benchmark on a TI TMS320C6678 8-core PDSP device.

3.5.1 Simulation

The mp-sched application involves mainly two actors: finite impulse response (FIR) filtering and fast Fourier transform (FFT) computation, which are fundamental operations in SDR applications.

Our FIR filter actor is specified using a single input FIFO, a single output FIFO, and one mode, the `process` mode. In this mode, the core operation of an FIR filter is developed in terms of the C language. The functionality is developed from the following equation, which defines the output sequence $y[n]$ in terms of the input sequence $x[n]$:

$$
\begin{aligned}
y[n] &= b_0 x[n] + b_1 x[n-1] + \cdots + b_N x[x-M+1] \\
&= \sum_{i=0}^{M-1} b_i x[n-i],
\end{aligned}
\tag{3.5}
$$

where $x[n]$ is the input signal, $y[n]$ is the output signal, the b_is are the filter coefficients, and M is the filter order, which is set to 79 in our design.

An FFT is an efficient algorithm to compute the discrete Fourier transform (DFT) and its inverse. FFTs are of great importance in a wide variety of applications, from digital signal processing and solving partial differential equations to algorithms for fast multiplication of large integers. Let $x_0, x_1, \ldots, x_{N-1}$ be complex numbers. The DFT is defined by the following formula:

$$
X_k = \sum_{n=0}^{N-1} x_n e^{-i2\pi k \frac{n}{N}}. \tag{3.6}
$$

There are N outputs $\{X_k\}$. Each output requires a sum of N items, which leads to $O(N^2)$ operations overall.

The most common form of FFT is the Cooley-Tukey algorithm, which calculates the DFT and its inverse using $O(N \log N)$ operations. This is a divide and conquer algorithm that recursively breaks down a DFT of any composite size $N = N_1 \times N_2$ into smaller DFTs of sizes N_1 and N_2, along with $O(N)$ multiplications by complex roots of unity. A radix-2 decimation-in-time (DIT) FFT is the simplest and most common form of the Cooley-Tukey algorithm. A Radix-2 DIT FFT computes the DFTs of the even-indexed inputs $x_{2m}(x_0, x_2, \ldots, x_{N-2})$ and of the odd-indexed inputs $x_{2m+1}(x_1, x_3, \ldots, x_{N-1})$, and then combines those two results to produce the DFT of the whole sequence. Then the same procedure is performed recursively to reduce the overall running time to $O(N \log N)$. The recursive tree of the Radix-2 DIT FFT is illustrated in Fig. 3.7a. Each white node in the figure represents a DFT computation involving the number of points annotated next to the node, and each black node merges two smaller DFT results together.

The FFT actor connects to one output FIFO and one input FIFO, and has one mode, the `process` mode. In this mode, we carry out the steps of the Radix-2 DIT FFT algorithm.

On each firing, the *SRC* node generates 1024 samples to each of its two output FIFOs. These blocks of 1024 samples are packaged into single tokens—i.e., each token encapsulates a complete block of 1024 samples. Such "blocking" of the data does not affect overall system input/output functionality, but it influences the internal dataflow structure of the design model and the associated analysis and implementation steps. The canonical scheduler is used to simulate the dataflow graph on a single core of 8-core PDSP. The processing time for the core computation (as shown in the dashed rectangle in Fig. 3.5) is 98.8 % of the computation time for the overall system. The overall computation time is 24.6 ms. The FFT computation takes 8.6 ms, while the FIR filter takes 3.55 ms. To improve the overall system performance, we need to optimize the core computation.

3.5.2 Implementation

In this section, we demonstrate, through the core computation of mp-sched, the use of both cross-platform implementation and scheduling/mapping in our proposed LWDF- and TDIF-based design flow.

3.5.2.1 Cross-Platform Implementation

To optimize the system, we first port the FIR filter and FFT actors to multiple PDSP cores by employing the multithreading libraries provided by TI. These libraries are provided with the TI PDSP platform to help DSP system designers express and exploit parallelism for efficient execution on the PDSP cores.

The FIR filtering operation provides a significant amount of data parallelism (DP). For demonstration purposes, this "intra-actor" DP is exploited across two cores on the targeted multicore PDSP using multithreading techniques. Our implementation can be readily adapted to utilize additional cores if desired.

Operation of our TDIF based FIR filter design is shown in Fig. 3.6, where X is single input token containing N input samples, Y is the output token, and B is the coefficient vector. In order to exploit DP, blocks of N input samples are divided

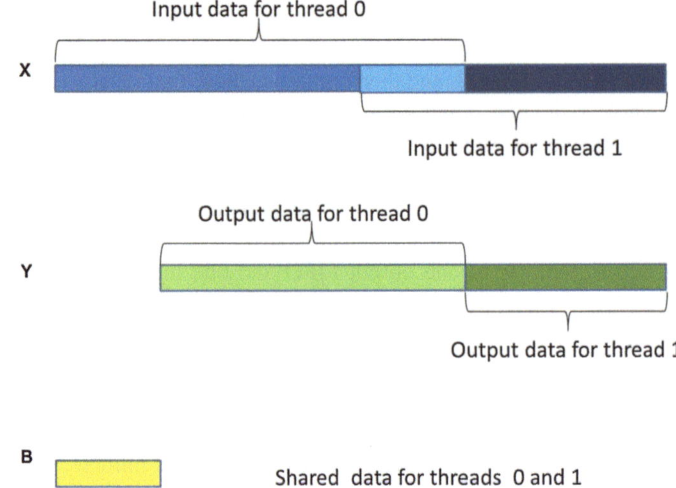

Fig. 3.6 Multithreaded FIR filter for PDSP implementation

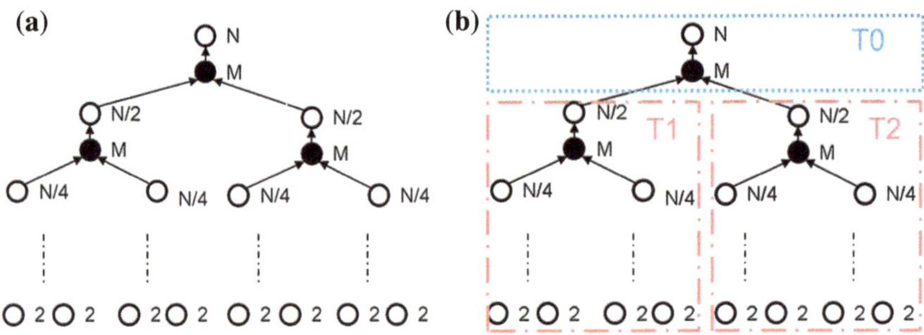

Fig. 3.7 Parallel FFT actor implementation using TDIF. **a** Recursive tree for Radix-2 DIT FFT. **b** Thread construction on recursive tree

into two groups, each for execution across two threads. Blocks of output samples are similarly divided into two groups each. The common coefficient vector B is shared by both threads, and each thread executes the `fir` calculation independently. This provides a multi-threaded FIR filter implementation, which exploits intra-actor parallelism in a manner that can be integrated systematically, through TDIF, with higher level parallelism exposed by the application dataflow graph. The overlap on the inputs of thread 0 and thread 1 arises because the current output depends on the previous $(M - 1)$ input samples.

To demonstrate the performance gain from multithreading, the execution time of a sequential 79th order FIR filter (Seq-FIR) design in simulation is compared to that of our parallel FIR (Par-FIR) implementation using identical input streams on the targeted multicore PDSP platform. The results for multiple input sizes are

Table 3.3 Execution time comparison between sequential FIR in simulation and parallel FIR implementations for different input sizes

Input size	1079	10079	100079	1000079
Seq-FIR (s)	0.0036	0.0336	0.334	3.34
Par-FIR (s)	0.0017	0.015	0.147	1.47
Speedup	2.11	2.24	2.27	2.27

demonstrated in Table 3.3. The input size is the number of signal samples. The reported execution time is the processing time, excluding the time for reading from the input FIFO and writing to the output FIFO. The FIFO reading and writing operations involve only pointer manipulations and no actual data movement, and thus have negligible impact on actor performance. The speedup is defined as the ratio between the execution time of Par-FIR and that of Seq-FIR. The super-linear speedup observed is due to the VLIW feature of the targeted kind of PDSP core. As more data is processed, more ILP is available to be exploited within each core.

DP and temporal parallelism (TP) from the recursive tree of Fig. 3.7a are utilized for multithreading in the FFT actor, as illustrated in Fig. 3.7b. With DP, two threads *T1* and *T2* each calculate half of the overall $DFT(N)$, which is $DFT(N/2)$ using the Cooley-Turkey algorithm. Another single thread *T0* merges the two results together. We experiment with two different implementations, I_a and I_b. In Implementation I_a, *T1* and *T2* are assigned to two separate cores, and *T0* is assigned to one of these two cores. In this implementation, only DP is exploited. On the other hand, in Implementation I_b, *T1* and *T2* are assigned to two separate cores, as in I_a, but *T0* is assigned to another (third) core. In this way, *T1* and *T2* can execute concurrently in a software pipelined manner with *T0* in a separate "pipeline stage". Implementation I_b thus exploits both DP and TP.

The execution time of the sequential FFT (Seq-FFT) in simulation is compared to the two different parallel FFT implementations, I_a and I_b. The execution time of I_a is the time for its longest pipeline stage. The speedup is defined by the ratio between the execution time of the parallel implementation (I_a or I_b) and that of Seq-FFT. The latency is another important figure of merit. The latency in these experiments is defined by the time difference $(t_w - t_r)$, where t_r denotes the time when the FFT actor reads the first token from its input FIFO, and t_w denotes the time when the actor writes the first result token to its output FIFO. The results are shown in Table 3.4. For Seq-FFT and I_a, the latency is the same as the execution time, so the latency values are not shown separately. We see from the results that I_b achieves more speedup than I_a, but introduces more latency.

3.5.2.2 Scheduling and Mapping

Using the GST representation for dataflow graph schedules, we have derived, by hand, three different scheduling and mapping schemes for the core computation of the mp-sched benchmark, and we have implemented these different schemes using the TDIF

Table 3.4 Execution time and latency comparisons between Seq-FFT in simulation, and the two parallel FFT implementations I_a and I_b

Input size	64	256	1024	4096
Seq-FFT (s)	0.00028	0.0016	0.0086	0.045
I_a (s)	0.00021	0.001	0.005	0.255
Speedup	1.3	1.61	1.72	1.788
I_b (s)	0.00023	0.00073	0.0039	0.021
Speedup	1.21	2.19	2.20	2.14
Latency (s)	0.00038	0.00118	0.00517	0.0257

scheduling APIs. Due to the use of a common underlying CFDF-based modeling foundation, the actor design from LWDF can be scheduled with no change in the TDIF scheduling APIs. Our experiments with these different scheduling and mapping schemes demonstrate the utility of our design flow for assessing design trade-offs based on alternative implementation strategies for a given application dataflow graph.

Note that the two paths in the graph of Fig. 3.5 are independent so that control parallelism can be used. Additionally, in each path, the individual actors can be pipelined to exploit more temporal parallelism. In the first scheme, we use a sequential implementation for each actor; however, we distribute different actors across different PDSP cores. In particular, the actors are mapped onto 4 DSP cores—*FIR1* is assigned to *DSP0*; *FFT1* is assigned to *DSP1*; *FFT2* is assigned to *DSP2*; and *FIR1* is assigned to *DSP3*.

The second and third schemes in our experimentation are derived by replacing sequential actor implementations in the graph with corresponding parallel actor implementations. In the second scheme, *FIR1* is assigned to *DSP0*, *DSP1*; *FFT1* is assigned to *DSP2*, *DSP3*; *FFT2* is assigned to *DSP4*, *DSP5*; and *FIR1* is assigned to *DSP6*, *DSP7*. On the other hand, in the third schedule, *FIR1* is assigned to *DSP0*; *FFT1* is assigned to *DSP1*, *DSP2*, *DSP3*; *FFT2* is assigned to *DSP4*, *DSP5*, *DSP6*; and *FIR1* is assigned to *DSP7*. Thus, in the second scheme, intra-actor parallelism is exploited for all actors, while in the third scheme, intra-actor parallelism is exploited only for the FFT actors. The GSTs with the associated actor and FIFO implementations are processed to generate corresponding application implementations for the targeted PDSP platform.

We experiment with three different implementations, corresponding to the three different schemes described above. The implementations are evaluated using the same input stream used in simulation, which contains *1024* samples. The execution time, speedup and latency values for these implementations are compared on the core computation shown in Fig. 3.5. The remaining actors (*SRC* and *SNK*) take only approximately 1.2 % of the computation time for sequential execution and thus do not have a significant impact on overall performance. The execution time is taken to be the processing time in the core computation region defined above. If pipelining is used, then the execution time is the time for the longest pipeline stage. The latency is defined as the elapsed time during the first iteration of graph execution between when the first input token enters the region and the time when the first output token leaves

Table 3.5 Execution time and latency comparisons among 3 different scheduling and mapping schemes and simulation for the mp-sched benchmark

Schedule	Execution time (s)	Speedup	Latency (s)
Simulation	0.0243	1	0.0243
1	0.00878	2.77	0.0089
2	0.00517	4.7	0.0052
3	0.0041	5.9	0.0092

the region. The results of the three implementations are compared to the simulation, as shown in Table 3.5.

The results demonstrate the capabilities in our design flow for supporting flexible design exploration in terms of different implementation constraints. For example, Scheme 2 has higher execution time but less latency compared to Scheme 3. If the system has a tight latency constraint, then Scheme 2 should be chosen, whereas Scheme 3 is preferable if throughput is the most critical constraint. Using our design flow, the designer can experiment with such alternatives to efficiently arrive at an implementation whose trade-offs are well matched to the design requirements.

3.6 Summary

In this chapter, we have introduced dataflow-based methods that facilitate software simulation and implementation for digital signal processing (DSP) applications. Specifically, we have demonstrated use of a design flow, based on core functional dataflow (CFDF) graphs, for simulation and implementation of diverse DSP systems—(1) an image processing application on a CPU/GPU platform, and (2) a software defined radio benchmark on a multicore programmable digital signal processor (PDSP).

Through these case studies on diverse platforms, we show that our design flow allows a designer to formulate and simulate DSP systems efficiently using lightweight dataflow (LWDF) programming. Then, from the profiling results in the simulation, the designer can identify bottlenecks in the system. To alleviate these bottlenecks, optimization techniques in the targeted dataflow interchange format (TDIF) are applied. While TDIF provides a framework where platform-specific optimizations can be applied and experimented with in a flexible way, our design methodology ensures that such experimentation adheres to the high level application structure defined by the simulation model. Such consistency is maintained as a natural by-product of the common CFDF modeling foundation that supports both LWDF and TDIF. The flexibility with which designers can implement different mapping and scheduling strategies using our design flow, and the efficiency with which such strategies can be integrated into complete implementations further facilitate exploration of optimization trade-offs.

Acknowledgments This research was sponsored in part by the Laboratory for Telecommunications Sciences, Texas Instruments, and US Air Force Research Laboratory.

References

1. Bhattacharyya, S.S., Deprettere, E., Leupers, R., Takala, J. (eds.): Handbook of Signal Processing Systems. Springer, Berlin (2010)
2. Lee, E.A., Messerschmitt, D.G.: Synchronous data flow. Proc. IEEE **75**(9), 1235–1245 (1987)
3. Lee, E.A., Parks, T.M.: Dataflow process networks. Proc. IEEE **83**(5), 773–799 (1995)
4. Plishker, W., Sane, N., Kiemb, M., Anand, K., Bhattacharyya, S.S.: Functional DIF for rapid prototyping. In: Proceedings of the International Symposium on Rapid System Prototyping, pp. 17–23. Monterey, California (2008)
5. Shen, C., Plishker, W., Wu, H., Bhattacharyya, S.S.: A lightweight dataflow approach for design and implementation of SDR systems. In: Proceedings of the Wireless Innovation Conference and Product Exposition, pp. 640–645. Washington DC, USA (2010)
6. Shen, C., Wu, S., Sane, N., Wu, H., Plishker, W., Bhattacharyya, S.S.: Design and synthesis for multimedia systems using the targeted dataflow interchange format. IEEE Trans. Multimedia **14**(3), 630–640 (2012)
7. Shen, C., Plishker, W., Bhattacharyya, S.S.: Dataflow-based design and implementation of image processing applications. In: Guan, L., He, Y., Kung, S. (eds.) Multimedia Image and Video Processing, 2nd edn., pp. 609–629. CRC Press, Boca Raton (2012). Chapter 24
8. Plishker, W., Sane, N., Kiemb, M., Bhattacharyya, S.S.: Heterogeneous design in functional DIF. In: Stenström, P. (ed.) Transactions on High-Performance Embedded Architectures and Compilers IV, Lecture Notes in Computer Science, vol. 6760, pp. 391–408. Springer, Berlin/Heidelberg (2011)
9. NVIDIA: NVIDIA CUDA Compute Unified Device Architecture: Programming Guide, Version 1.0 (2007)
10. Texas Instruments, Inc.: TMS320C6678 Multicore Fixed and Floating-Point Digital Signal Processor Data Manual (2012)
11. Ko, M., Zissulescu, C., Puthenpurayil, S., Bhattacharyya, S.S., Kienhuis, B., Deprettere, E.: Parameterized looped schedules for compact representation of execution sequences in DSP hardware and software implementation. IEEE Trans. Signal Process. **55**(6), 3126–3138 (2007)
12. Plishker, W., Sane, N., Bhattacharyya, S.S.: A generalized scheduling approach for dynamic dataflow applications. In: Proceedings of the Design, Automation and Test in Europe Conference and Exhibition, pp. 111–116. Nice, France (2009)
13. Bhattacharyya, S.S., Murthy, P.K., Lee, E.A.: APGAN and RPMC: Complementary heuristics for translating DSP block diagrams into efficient software implementations. J. Des. Autom. Embed. Syst. **2**(1), 33–60 (1997)
14. Wu, S., Shen, C., Sane, N., Davis, K., Bhattacharyya, S.: Parameterized scheduling for signal processing systems using topological patterns. In: Proceedings of the International Conference on Acoustics, Speech, and Signal Processing, pp. 1561–1564. Kyoto, Japan (2012)
15. Bilsen, G., Engels, M., Lauwereins, R., Peperstraete, J.A.: Cyclo-static dataflow. IEEE Trans. Signal Process. **44**(2), 397–408 (1996)
16. Plishker, W., Sane, N., Bhattacharyya, S.S.: Mode grouping for more effective generalized scheduling of dynamic dataflow applications. In: Proceedings of the Design Automation Conference, pp. 923–926. San Francisco (2009)
17. Owens, J.D., Houston, M., Luebke, D., Green, S., Stone, J.E., Phillips, J.C.: GPU computing. Proc. IEEE **96**(5), 879–899 (2008)
18. Ritz, S., Pankert, M., Meyr, H.: Optimum vectorization of scalable synchronous dataflow graphs. Proceedings of the International Conference on Application Specific Array Processors, In (1993)

19. Zivojnovic, V., Ritz, S., Meyr, H.: Retiming of DSP programs for optimum vectorization. In: Proceedings of the International Conference on Acoustics, Speech, and, Signal Processing, pp. 492–496 (1994)
20. Lalgudi, K.N., Papaefthymiou, M.C., Potkonjak, M.: Optimizing computations for effective block-processing. ACM Trans. Des. Autom. Electron. Syst. **5**(3), 604–630 (2000)
21. Ko, M., Shen, C., Bhattacharyya, S.S.: Memory-constrained block processing for DSP software optimization. J. Signal Process. Syst. **50**(2), 163–177 (2008)
22. Lapsley, P., Bier, J., Shoham, A., Lee, E.A.: DSP Processor Fundamentals. Berkeley Design Technology, Inc. (1994)
23. Zaki, G., Plishker, W., Bhattacharyya, S., Clancy, C., Kuykendall, J.: Vectorization and mapping of software defined radio applications on heterogeneous multi-processor platforms. In: Proceedings of the IEEE Workshop on Signal Processing Systems, pp. 31–36. Beirut, Lebanon (2011)
24. Blossom, E.: GNU radio: Tools for exploring the radio frequency spectrum. Linux J. **2004**(122), 4 (2004)

Part II
Model-Driven Design, Integration and Verification of Heterogeneous Models

Chapter 4
Model-Driven Design of Software Defined Radio Applications Based on UML

Jair Gonzalez and Renaud Pacalet

Abstract Model-driven design (MDD) is considered a very promising approach to cope with the design of complex software applications such as software defined radio (SDR). This chapter proposes an MDD methodology for SDR applications. Our approach comprises: (1) DiplodocusDF, a domain-specific modelling language for SDR applications, it is a domain specific UML profile. (2) The mechanism to transform DiplodocusDF models into C-language code ready for compilation, and (3) a runtime environment for execution of the generated code. Moreover, the proposed UML profile is supported by TTool, which is a framework for design exploration and formal verification at model level. We illustrate the potential of our methodology designing a SDR application.

4.1 Introduction

Current design methodologies cannot cope efficiently with the high complexity of software defined radio (SDR) applications [1]. SDR applications are extremely complex as they have to conciliate many different opposing factors, for example: high-performance and hard-real-time, inter-operability and safety, etc. Also, novel SDR platforms, such as expressMIMO [2], account for the software complexity, as they are tuned for computation efficiency, in terms of speed and power consumption, but its operation is hard and error-prone. It is compulsory to introduce a new design methodology where the execution platform is abstracted, allowing the domain expert to concentrate only on the pure application matters.

Jair Gonzalez (✉) · Renaud Pacalet
Telecom ParisTech, LTCI CNRS, 2229 Route des Cretes,
B. P. 193, Sophia-Antipolis Cedex, 06904 Paris, France
e-mail: Jair@telecom-paristech.fr

Renaud Pacalet
e-mail: Renaud.Pacalet@telecom-paristech.fr

A. Sangiovanni-Vincentelli et al. (eds.), *Embedded Systems Development*, Embedded Systems 20, DOI: 10.1007/978-1-4614-3879-3_4,

Model driven design (MDD)is considered to be a promising approach to handle the development of complex software systems [3–5]. MDD allows for describing the applications in the problem-space rather than in the solution-space. In MDD, the problem is described by the designer using an intuitive domain-specific modelling language, and the implementation solution is (ideally) found automatically by synthesis mechanisms. This releases the designer from having to be a platform and a low-level language expert. This sole abstraction simplifies the design effort and reduces the possibility of implementation errors.

This chapter describes a MDD methodology for the domain of SDR applications. It includes a domain-specific modelling language called DiplodocusDF, the synthesis mechanisms capable of transforming a DiplodocusDF model into C-language code, and an execution environment. DiplodocusDF is based on UML and dataflow modelling. dataflow modelling is a good candidate to describe SDR applications as it helps to exhibit the potential parallelism among SDR operations, and thus take profit from multiprocessing platforms such as expressMIMO [2] or others. UML was chosen given its capabilities to describe the application and the architecture independently, this permits the description of portable applications. UML allows incremental refinement and abstract simulation, therefore we can evaluate design choices in several steps. Also, the UML descriptions can be translated into formal ones for formal verification. Last but not least, UML poses robust and well accepted graphical notations and flexibility to be extended to cover domain-specific applications such as SDR. The approach is exemplified by designing a SDR application, which is modeled, simulated, mapped into a given architecture model, translated into executable C-code and executed. Our methodology is supported by TTool [6].

The rest of the chapter is organized as follows: the related work is described in Sect. 4.2. Section 4.3 discus with more details the MDD methodology proposed in this chapter. Section 4.4 describes the proposed modelling profile DiplodocusDF. Section 4.5.2 explains briefly the transformation mechanisms from DiplodocusDF to C-language. Section 4.6 elaborates on the runtime environment that supports the execution of the generated applications. A case of study is presented in Sect. 4.7. The future work and conclusions are discussed in Sect. 4.8.

4.2 Related Work

Software defined radio are in fact dataflow applications. There exist already some languages to describe them, for example StreamIt [7], which is a textual language for dataflow applications to be executed in symmetric multicore platforms (SMP). While there exist some SMP SDR platforms, the bast majority have an heterogeneous architecture, making the StreamIt concepts unsuitable. Another language is the radio description language (RDL) [8], which is a domain specific language for SDR, they follow the same approach as us in dividing the processing functionality in two: in one side the SDR operations (a unit of digital signal processing, for example FFT), in the other side the SDR waveform which configures the DSP operations and triggers its execution.

Domain specific languages should have certain characteristics to be successfully adopted [9]. This characteristics can be achieved by using UML as description language [4], bib1231146. The authors of [10], bib1541072, and [12] propose methodologies to generate SDR operations from UML models. The author of [13] discusses the feasibility of mixing UML and SCA to create an hybrid methodology for the codesign of SDR applications. He suggests using UML RT to describe the SDR waveforms. Using UML RT, as it is meant for embedded control applications, results in complex models difficult to maintain and which constraint the solution space by providing a fixed scheduling of SDR operations, it is necessary to extend UML to cover dataflow applications [14]. The paper [15] presents an approach to use activity diagrams of UML 2.0 for dataflow applications. Their proposal is similar to ours in covering flat control-flow and dataflow, as suggested in [16], but their objective is only modelling and no concepts of synthesis are taken into account.

The contribution of this work is the definition and implementation of a methodology for SDR applications, which covers not only the performance but also waveform portability requirements. Portability means that we can use the same waveform description for different platforms, for example, a cell phone and a base station. We achieve portability by two means: **First**, we divided the SDR applications in two parts: The SDR operations and the SDR waveforms. The SDR operations represent the actual signal processing mechanisms and are modeled as components of a given platform. The SDR waveforms are coordination applications that trigger and configure the execution of the SDR operations. **Second**, by introducing an extra layer to the execution platform called runtime environment. This layer interprets the waveforms and resolves its intra/inter data dependencies. Although the runtime environment introduces some processing overhead, its benefits are more important, as it also helps to satisfy the field reconfigurability and system multimodal requirements.

4.3 Proposed Design Methodology

We propose a model driven design methodology (depicted in Fig. 4.1) for software defined radio applications. It is composed of the following steps:

1. **Abstract modelling** of waveforms and architectures independently one from the other.
2. **Mapping** the waveforms to the selected architecture.
3. **Formal verification** by transforming the mapped application into a formal representation to validate liveness properties.
4. **Code generation** to systemC or c-language. To generate C-language, it is necessary an extended description of the target architecture, called Model extension constructs (MEC), which provides the constructions that are relevant for code generation but not for simulation.
5. **Cycle-accurate simulation** based on the generated systemC code, required to validate performance requirements at model level.

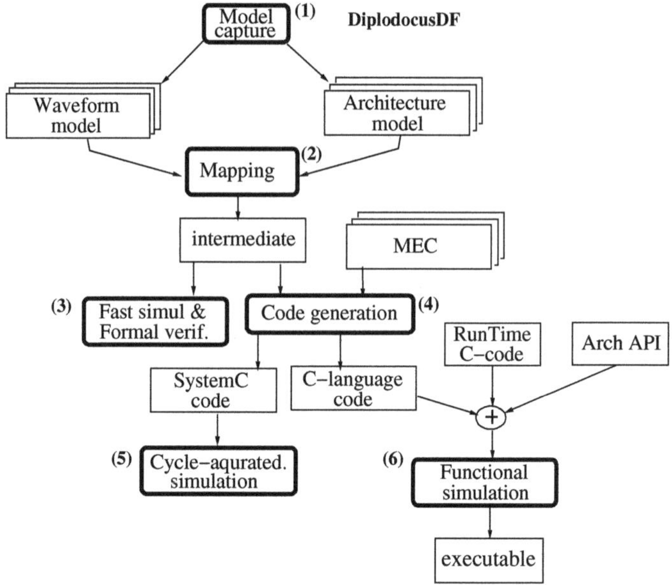

Fig. 4.1 Proposed model-driven design methodology

6. **Functional simulation** based on the generated C-language code and the an API which describes the function calls to the SDR operations supported by the given architecture.

Steps (1), (2), (3) and (6) have been integrated into current design flows, but as noted in [3, 5], there remains a synchrony problem between the model and the final implementation, i.e., the final implementations does not reflect the initial model. To solve this, MDD methodology includes step (4), which is the automatic generation of code based on the model.

In this chapter we concentrate on steps (1), (2), and (4). The synthesis step corresponds only to code translation. Further development of the synthesis mechanism will include automatic search of memory allocation and scheduling solutions.

4.4 DiplodocusDF

DiplodocusDF is a UML profile specific for SDR applications. A key component of a MDD methodology is the domain-specific modelling language (DSML) that captures accurately the semantics of SDR applications. A DSML should meet certain criteria in order to get to be effective. First, it should differentiate the pure application aspects from the implementation choices, i.e., the models should avoid (as much as possible) the implementation constraints, as they reduce the solution-space. Second, it should be complete in the sense that it captures all the information that is necessary to generate executable code. Other aspects to consider when defining a DSML are: it should be an intuitive language for the domain expert and it should be predictive, that is, it should be evident what the result will be after automatic code generation.

DiplodocusDF follows a declarative/dataflow approach. This allows describing the pure applications aspects, without considering the implementation details. The declarative approach contrasts to the procedural approach (followed by regular UML/MARTE profiles) in that declarative describes what the solution is, while the procedural approach describes how to find the solution. Procedural approaches lead to models where one component is on charge of controlling the execution of the rest. That controlling component represents in fact scheduling design choices, which limit the solution space. DiplodocusDF describes only data dependencies between process operators, it does not constrain the order of execution of two non-dependent operators.

DiplodocusDF is in a high degree similar to Kahn process networks [17] (KPN), KPN are appropriate for SDR as it enforces portability, determinism and parallelism [16], although Kahn networks imply unbounded memory, this is not true for real target architectures. Therefore the execution of operations hast to be constrained to limit memory usage, an operator cannot be executed if its previously generated data has not been consumed. The resulting execution is still a valid Kahn execution, holding the property of determinism, i.e., for the same inputs, it always produces the same outputs.

4.4.1 SDR Waveform Notations

DiplodocusDF is based on three model notations to create SDR waveform models: (1) Dataflow operators (2) Dataflow ports and (3) Dataflow links. They extend, in the same order, the component, port and link concepts of DIPLODOCUS to support the pair (data-block, clock) of synchronous dataflow models.

Dataflow ports are the interface to exchange *dataflow* between dataflow operators. They support dataflow semantics, i.e., data-block and clock transference. For this they merge two types of ports supported in DIPLODOCUS: channels and events. The channel part models the data-block component of dataflow, while the event part models the clock component. The event signals the presence of a valid data-block in the channel. The channel is configured to be non-blocking read/non-blocking write (NBR/NBW), allowing dynamic data-block size during simulation, therefore component re-utilization. Each dataflow port can be linked to only one other port. A dataflow operator can have one or more input and output ports.

Dataflow links are used to describe data-block dependencies among the dataflow operators, data is sent/received from/to an operator through its dataflow ports, links only describe dependency.

Dataflow operators there are two types of operators: execution operators and router operators. The execution operators transform the data-block values while the router operators transform the sequence of the values in the data-block. For example, a component-wise addition operator will change the values of the input data, while an overlapping operator will not change the values but its sequence. This approach permits defining reusable model operators, which are instantiated and configured by the user. There are two subcomponents for the operators, DBox and FBox. DBox describes the actual execution of data transformation, while FBox describes the dataflow. The execution operators are composed both, FBox and DBox. While the router operators require only FBox (as data is not transformed by these operators). The FBox waits for input clock signal, generates the configuration parameters, fires the corresponding DBox, and sends a clock event through the output dataflow port(s). The parameters of execution are sent to the DBox through an internal request port. The process in a DBox involves: reading the data-block from the input dataflow port(s), execute the operation, and sending the resulting data-block though the NBR/NBW output port.

The behaviour of both sub-components is described using activity diagrams. The Fig. 4.2 shows two operators, `OVLP` and `FFT` and its activity diagrams, **F_fft** and **X_fft** for `FFT`, and **OVLP** for `OVLP`, as `OVLP` is a router operator, it does not require DBox.

The behaviour of `FFT` is as follows: It waits in **F_fft** for the arrival of a clock event in the `FFT_in` port, the event should come together with the parameter `fft_size`. After having received the clock event, **F_fft** request the execution of **X_fft** through `r_fft` and blocks. Then the **X_fft** is executed, it reads from `FFT_in` the data-block of size `fft_size`, the processing is modeled by `I`, then the resulting data is sent through `FFT_out` and the execution returns to **F_fft**, where an clock event is sent through `FFT_out` with the size `fft_size`, but before the event `FFT_in` is cleared to indicate that the input data was consumed.

4.4.2 Target Architecture Notations and Mapping

As mentioned before, in DiplodocusDF, the application model is described independently of the architecture model. The architecture profile of DiplodocusDF allows

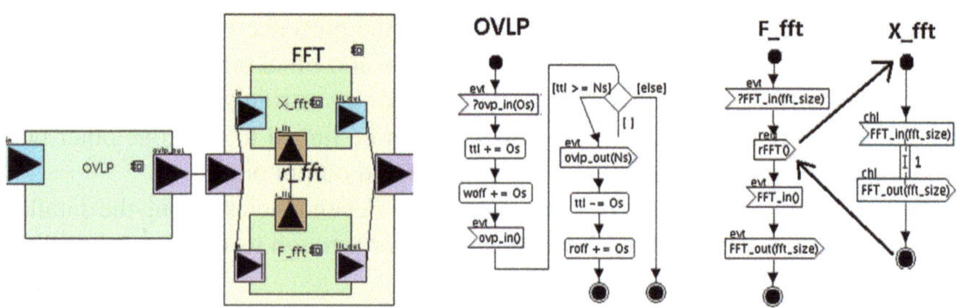

Fig. 4.2 Router (OVLP) and execution (FFT) operators and its activity diagrams

describing a wide range of target architectures, which can be based on a single processor, or multiple processors, either homogeneous or heterogeneous, combined with a large variety of memory schemes. DiplodocusDF architecture profile is based on three main notations:

Data processing nodes can be of two types: general-purpose or specific-purpose. All FBoxes have to be mapped to general-purpose processors, while the DBox can be mapped to both types. The difference between both types is that general-purpose processing nodes implement a sharing policy to execute two or more tasks simultaneously, while the specific-purpose processing nodes execute one task at a time until completion. Fig. 4.5 shows an architecture with three processing nodes $\ll CPURR \gg$, `FEP`, `DMA`, and `LEON`. The Fig. 4.5 also show the mapping of all DBoxes and FBoxes of the waveform described in Sect. 4.7.

Data storage nodes are used to describe the memory components. They are composed of buffer blocks, which is has two basic parameters: base address and size. Dataflow links are mapped to buffers statically by the user. Later the synthesis mechanism will be improved in such a way that the buffer configurations (base address and size) and mappings will be generated automatically. The Fig. 4.5 shows two storage nodes $\ll MEMORY \gg$, `FEP_MSS` and `main`.

Data transference nodes can be of types: bus and bridge. Bus nodes are placed between two processing nodes or one processing node and one storage node. They describe data accessibility, for example node `FEP` from Fig. 4.5 can share data with nodes `FEP_MSS` and `DMA`, but not with nodes `LEON` or `main`. Each bus has a particular arbitration policy. Bridges are meant to communicate two buses.

4.4.3 Performance Requirements Notations

SDR applications have soft/hard real-time requirements. For example, LTE has synchronization information that has to be processed periodically. If this data is not processed on time, the terminal will loose synchronization. The SDR waveform carries required processing time (RPT). This information is complemented by the SDR platform model, which has information on how much time it takes to process an unit of information. At mapping time we know how much information will be processed by each SDR platform element, and therefore how much time this processing element will take to process the incoming signal, we call this platform processing time (PPT). The RPT and PPT are used to calculate a dynamic priority for each SDR operation available for execution.

4.5 Code Generation

The execution code is generated from the intermediate description (ID) (see Fig. 4.1). The ID includes information from the waveform model, architecture model, mapping and Model Extension Constructs (MEC). The ID is translated to C-language by our code generation mechanism.

4.5.1 Model Extension Constructs

The MECs describe the particular function call constructions for a given operation/architecture. While the application describes *what* function will be called, the MECs add information on *how* the function calls look like. For example, the FBox `F_fft` of operator `FFT` shown in Fig. 4.2, request the execution of the DBox though the `rFFT` request, this is noted in the intermediate format as $REQUESTrFFT$, which is architecture independent notation. The mapping information shows that the requested DBox is assigned to the `FFT` processing node, which has an associated MEC. The associated MEC for `FFT` operation is $fep\_start(\&XOP)$, where XOP is replaced by X_fft (name of the DBox of `FFT`), as shown in the generated code, Fig. 4.6 line 33.

4.5.2 DiplodocusDF Translation Semantics

Operators The model operators are translated to C as coordination functions (CF) which are executed when their input data is available. The attributes of the UML operator model are translated as variables and the activity diagram of the FBox is translated to C-code. The logic of the code is obtained from the activity diagrams, while the particular constructions are obtained from the model extension constructions associated to a given architecture.

Dataflow ports A dataflow port is an abstract representation of an operand, including its name and its memory location. Writing to a dataflow port means writing to a memory location, same for reading. A dataflow port is translated as an instance of a dataflow signal class. This class has different attributes, some depend on the target platform and some are common to all the platforms. The common attributes are: (1) a flag to indicate when the data-block is valid, (2) the length of the data-block, and (3) the block number. The block number is used by the routing operators to take decisions on how to deal with that block. If the target platform was expressMIMO [2], then the dataflow signal class includes: (1) a base address, (2) a bank number, and (3) the type of data to be stored. If the platform was a PC, it would only be required a pointer to memory. Other platforms might require different attributes to describe the location of the operands.

Dataflow links Represent dataflow dependencies among operators. After mapping once data-transference operators were inferred, When the ports connected by a dataflow link have the same name, they refer to the same signal. Then the link inherits that name and it is shown in the model as such.

4.6 Runtime Environment

The applications are executed with the support of a runtime environment. The functioning of the environment is depicted in Fig. 4.3, it shows a configuration made of n applications that are executed on a $m + 1$ processors platform. At least one of the processors should be a general purpose, it is on charge of executing the coordination applications, the rest of the processors (potentially specialized) execute the SDR operations.

The runtime environment architecture follows a data-driven approach. In a data-driven approach, the functions are executed when its input data is available. It was said before that a DiplodocusDF model is translated into a coordination application. In the proposed Runtime environment, The coordination applications are composed of a scheduler, a set of coordination functions (list CF in Fig. 4.3), one for each operator in the model, and a set of signals (list S in Fig. 4.3), one for each dataflow link. The scheduler executes the coordination functions when its input(s) are valid and its output(s) are marked as consumed. This is similar to what happens at model level. It was chosen in this way in order to maintain the determinism and parallelism properties as described by the model. It also allows to generate regular code, predictable from the model, as suggested in [3] for effective model-driven design.

The coordination function manages the parameters of the SDR operations, and sets them for execution by aggregating them into the execution queue of a co-processor.

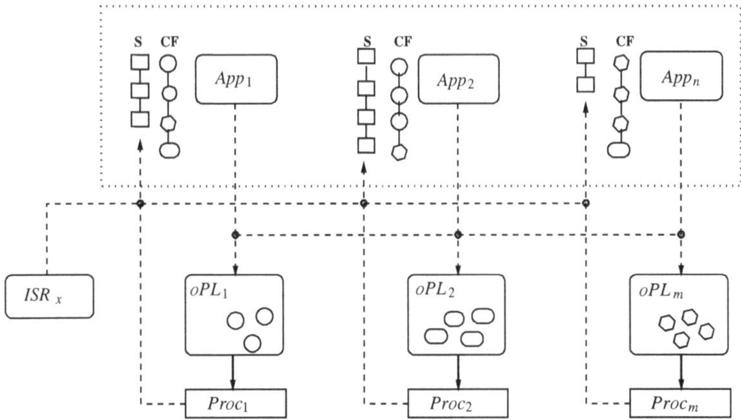

Fig. 4.3 Runtime environment

In Fig. 4.3, there are m ordered priority lists (oPL_m) which hold the operations from all the applications in the configuration. The ordered list is reordered every time a new operation is pushed into, such a way that a pop from the list will return the operator with the highest priority. The idea is to keep the co-processors in constant execution.

The scheduling of the coordination applications is done in a cooperative way, similar to the time-triggered hybrid (TTH) approach proposed in [18], where a set of software applications are executed according to a schedule defined before of execution. The applications cannot preempt each other but can be preemted by a third type of tasks that are interrupt dependent, for example, an interrupt coming from a analog-to-digital converter, signalling the reception of data. Another source of interruption can be an specialized co-processor signalling that it is ready to execute another SDR operation. The interruption routines should be very quick, in such a way that its execution does not deviate significantly the static scheduling of the coordination applications.

In the coordination functions, there are two main classes of objects, the coordination functions (CF) and the signals (S). Each CF object corresponds to one of the components from its DiplodocusDF model. Each S object in the system corresponds to one of the dataflow links. The CF class has two properties, CF id, and CR priority. The Signal class is more complex, it does not carry the actual data but its location in memory. This is platform dependent information that is added to the model by the user at mapping time. For the case of expressMIMO, the signal parameters are: base address, length of data vector, data type, memory bank, and memory block. It also includes a valid flag to indicate when the data is ready to be read. The Operator class has four methods: the constructor, the execution, the fire rule, and the destructor. The constructor method initializes the parameters of its corresponding output signal(s). That means that the operators know in advance where in memory they will write their results. This information is captured in the model by the user at mapping time. There is also the execution method, which implements a c-language version of its activity diagram from the SDR waveform model. The method is implemented in a run-to-completion approach, i.e., the method should not block. The fire rule method is used to evaluate the signals condition before firing the execution method. For example, while the `A` component is executed whenever one of its inputs is valid, the `CWA` component requires both input signals to be present.

4.7 DiplodocusDF example: Welch Periodogram Detector

This section demonstrates the use of DiplodocusDF by describing the Welch periodogram application for opportunistic spectrum sensing, recently proposed in [19]. Welch periodogram helps to identify the significance of the frequency contributors in a signal by calculating sub-band's energy, and comparing to its designated threshold. The algorithm is described in Algorithm 4.1 and the model is shown in Fig. 4.4. The application was mapped into the target architecture of Fig. 4.5.

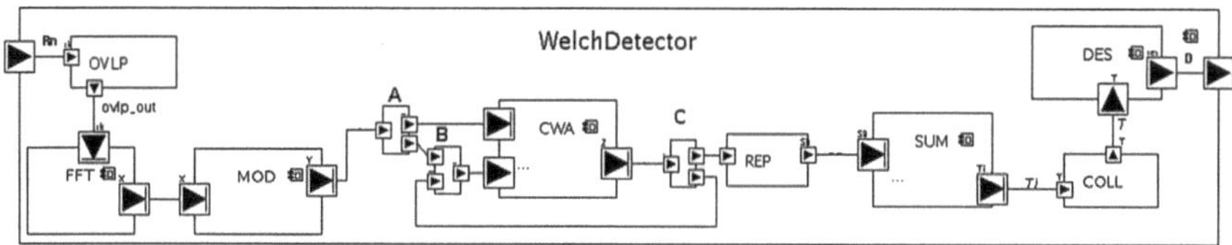

Fig. 4.4 DiplodocusDF application model of Welch detector

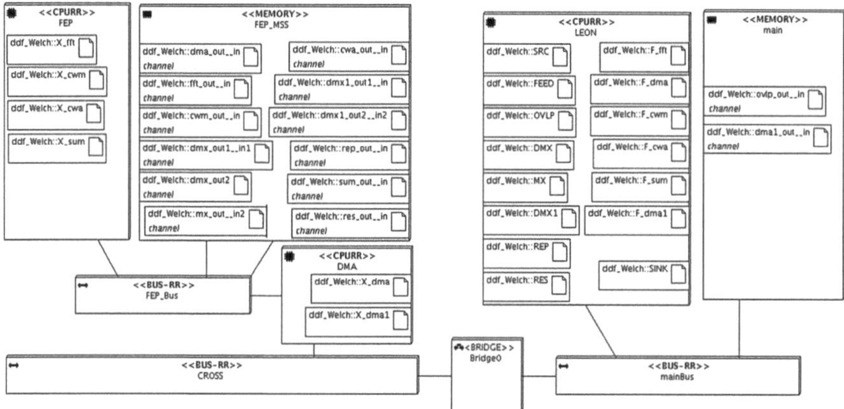

Fig. 4.5 DiplodocusDF architecture model and mapping of Welch periodogram application

---

Algorithm 1: Welch detector algorithm

---

input : r_k, m, N_s, O_s, data block, Number of blocks, block size, and block overlapp size .
input : I, Number of sub-bands (SBs) of interest.
input : f_i, L_i, Y_i, T_i, D_i, Lists SB's central frequencies, widths, thresholds, statistics, and decisions.
temporal: X, Y, Z, R_s, Buffers of size N_s.

1 $t \leftarrow N_s - O_s$;
2 **for** $k \leftarrow 0$ **to** $m - 1$ **do**
3 $\quad r_k \leftarrow [r(kt), r(1 + kt), \ldots, r(kt + N_s + 1)]$;
4 $\quad X \leftarrow$ FFT(N_s, r_k);
5 $\quad Y \leftarrow$ MOD(N_s, X);
6 $\quad Z \leftarrow$ CWA(N_s, Z, Y);
7 **end**
8 **for** $i \leftarrow 1$ **to** I **do**
9 $\quad start \leftarrow f_i - \frac{L_i}{2}$;
10 $\quad T_i \leftarrow$ SUM$(L_i, start, Z)$;
11 $\quad$ **if** $T_i > Y_i$ **then**
12 $\quad\quad D_i \leftarrow 1$;
13 $\quad$ **else**
14 $\quad\quad D_i \leftarrow 0$;
15 $\quad$ **end**
16 **end**

Model capture The line 3 in Algorithm 4.1 describes the overlapping of block elements. This operation is modelled in Fig. 4.4 by component OVLP. The behaviour of OVLP is described by the activity diagram **OVLP** of Fig. 4.2 . It simply waits for the arrival of data-blocks at its input port (ovlp_in), whenever N_s elements are available, they are sent to the next component through the output port (ovlp_out). After sending, the first O_s received elements are discarded. The size O_s of the received data-block is received together with the data-block.

The line 4 corresponds to the component `FFT` in the model. Here the input data is transformed to the frequency domain. The line 5 corresponds to `MOD` component in the model, there the magnitude of each element of the input block is calculated. Line 6 accumulates the magnitude of each element of the block, this corresponds to `CWA` in the model. These three execution operations have similar behaviours, they wait for the input signals, after receiving them they are operated and the output is sent through the output port. In the case of `CWA`, it waits for the two input signals and they have to be present before proceeding. The difference between the operators is more important at runtime, described in Sect. 4.6.

The components `A`, `B` and `C` are used to construct the dataflow loop of line 2. The `A` component is described in such a way that the first block is routed to the output port `out2` (the lower one), the rest of the blocks are routed to the output port `out1` (upper one). It is important to note that each data-block carries its data-block identifier. The component `B` is described to retransmit from input port `in1` only the first block, its port `in2` is always retransmitted. These components are executed when at least one of the input signals is available. Component `C` is described in such a way that the input is retransmitted through output `out1` only when it is the number $m - 1$, the block is retransmitted through `out2` in any other case.

The execution operator `SUM` corresponds to line 10, it adds the elements of the received block and sends the one-element result block through its output. The `REP` receives one full block of N_s elements and sends L sub-blocks of its corresponding L_i size. The parameters (L, L_i, f_{off}^i) are also received as part of the input block information. The `COLL` dataflow operator receives m one-element blocks and sends one block of m elements. The `DES` operator receives the block and compares each of its elements to its corresponding threshold $Y_w d$. `DES` generates a block of L elements containing the decisions for each element (sub-band). This corresponds to the output of the Welch periodogram model, which will activate a next operator in the overall model.

Architecture and Mapping The application is mapped into the heterogeneous multiprocessor architecture, expressMIMO [2]. The architecture model of Fig. 4.5 shows only the front-end processing shell, composed of an IP-core (`FEP`), memory sub-system(`FEP_MSS`), and DMA engine (`DMA`). It also shows the general purpose processor (`LEON`), which is on charge of executing the FBoxes. The DBoxes where mapped to either `FEP` processing node or the `DMA` processing node.

Code translation The Fig. 4.6 shows part of the final C-code generated automatically. Line 3 and 22 show two coordination functions (CFs), op_OVLP is a router function and op_F_fft is the CF corresponding to the FBox of the `FFT` operator. Lines 43 and 47 show the corresponding fire rule functions. The structure of at line 55 shows a MEC corresponding to the expressMIMO platform, if the application was mapped to a different platform, the BUF_TYPE declaration would follow the particularities of the corresponding target. The list of operations(ops_enu) and links ($sigs\_enu$) are shown in lines 78 and 72 respectively. A very simple scheduling

```
1 /******************************/
2 /** Coordination functions ***/
3 void op_OVLP(){
4 int status=0;
5 static int in_size, valid;
6 static int roff=0;
7 static int woff=0;
8 static int Ns=40;
9 static int Os=20;
10 valid=valid+in_size;
11 woff=woff+in_size;
12  if (valid>=Ns){
13   SIG[ovlp_out].pBuff=SIG[dma_out].pBuff;
14   SIG[ovlp_out].flag=true;
15   SIG[dma_out].flag=false;
16   valid=valid-Os;
17   roff=roff+Os;
18  }
19 return status;
20 }
21
22 void op_F_fft(){
23 int status=0;
24 /*firm configuration*/
25  fep_set_l(&X_fft, sig[ovlp_out].l);
26  fep_set_qx(&X_fft,sig[ovlp_out].pBuff->q);
27  fep_set_bx(&X_fft,sig[ovlp_out].pBuff->b);
28  fep_set_tx(&X_fft,sig[ovlp_out].pBuff->t);
29  fep_set_qz(&X_fft,sig[fft_out].pBuff->q);
30  fep_set_bz(&X_fft,sig[fft_out].pBuff->b);
31  fep_set_tz(&X_fft,sig[fft_out].pBuff->t);
32 /*start execution
33 */fep_start(&X_fft);
34
35 SIG[fft_out].flag=true;
36 SIG[ovlp_out].flag=false;
37 return status;
38 }
39
40
41 /******************************/
42 /**** Fire rule functions ****/
43 bool fr_OVLP(void){
44   return((sig[dma_out]) && (!sig[ovlp_out]));
45 }
46
47 bool fr_F_fft(void){
48   return((sig[ovlp_out]) && (!sig[fft_out]));
49 }
50
51
52
53 /******************************/
54 /*** struct definitions *****/
55 struct FEP_BUFF_TYPE{
56  int l; //sample length
57  int b; //sample base address
58  int q; //bank 0,1,2,3
59  int t; //type of data
60 };
61 typedef struct FEP_BUFF_TYPE FEP_BUFF_TYPE;
62
63 struct SIG_TYPE{
64  bool flag;  //new signal flag==1
65  int woff;  //write offset
66  int roff;  //read offset
67  B_TYPE*pBuff;
68 };
69 typedef struct SIG_TYPE SIG_TYPE;
70
71
72 enum sigs_enu{
73  cwa_out, cwm_out, dma1_out, dma_out,
74  dmx1_out1, dmx1_out2, dmx_out1, dmx_out2,
75 feed_out, fft_out, mx_out, ovlp_out, rep_out,
76 res_out, src_out, sum_out, NUM_SIGS};
77
78 enum ops_enu{
79  DMX, DMX1, FEED, F_cwa, F_cwm, F_dma, F_dma1,
80 F_fft, F_sum, MX, OVLP, REP, RES, NUM_OPS};
81
82 /******************************/
83 /*** run-time   **************/
84 int main(void){
85 bool valid_signal = false;
86 int status = 0;
87
88  register_operations();
89  register_fire_rules();
90  init_operations_context();
91
92  while(!sig[des_out]){
93   for(int n_op=0; n_op<NUM_OPS; ++n_op){
94    //evaluate the fire rule
95    valid_signal = (*fire_rule[n_op])();
96    if(valid_signal){
97     // execute coordination function if
98     // its fire rule is met
99     status = (*operation[n_op])();
100    }
101   }
102  }
103  clear_contexts();
104 }
```

Fig. 4.6 Extracts from the code that was generated automatically

implementation is shown in the main function at line 84. The code was executed with results that are functionally correct.

4.8 Conclusions

This work presents a model-driven design methodology for the design of software defined radio applications, which present the challenge of reconciliation of real-time requirements and portability requirements. We achieve this by dividing the SDR system in two: The SDR waveforms and the SDR operations provided by SDR platforms. We provide the UML profiles to describe separately the waveforms from the platforms. We also designed and implemented the mechanisms to transform the UML model into executable code. The transformation is supported by a a model extension language, which provides the specific language constructs related to the given platform. We designed and implemented a runtime environment, which gives a standard interface to the generated waveforms. The runtime allows portability and reconfigurability as the same waveform description can be executed by any platform equipped with the runtime. It also helps to retain the determinism properties of the SDR waveform model.

We demonstrated the use of our methodology, profiles, transformation mechanisms, and runtime, by designing an energy detection waveform. Energy detection is a fundamental component for SDR/cognitive radio applications, necessary to detect available radio spectrum. Welch periodogram has been proved [19] to be very effective for fast energy detection. We demonstrate the design and deployment on the expressMIMO platform [2].

Much remains to be done, until now, the memory allocation and scheduling solutions are provided by the user. we continue working in to provide the algorithms/mechanisms that will generate the buffers addressing automatically. Also, it is necessary to enhance the requirements and constraints profile to direct/generate correct scheduling solutions in terms of performance requirements.

Acknowledgments The research leading to these results has received funding from the European Community's Seventh Framework Programme (FP7/2007-2013) under grant agreement SACRA n° 249060.

References

1. Ulversoy, T.: Software defined radio: Challenges and opportunities. Communications surveys tutorials, IEEE PP(99), 1–20 (2010). doi:10.1109/SURV.2010.032910.00019
2. Nussbaum, D., Kalfallah, K., Knopp, R., Moy, C., Nafkha, A., Leray, P., Delorme, M., Palicot, J., Martin, J., Clermidy, F., Mercier, B., Pacalet, R.: Open platform for prototyping of advanced software defined radio and cognitive radio techniques. In: Digital system design, architectures, methods and tools, 2009. DSD '09. 12th euromicro conference on, pp. 435–440 (2009). doi:10.1109/DSD.2009.123
3. Schmidt, D.: Guest editor's introduction: Model-driven engineering. Computer **39**(2), 25–31 (2006). doi:10.1109/MC.2006.58
4. France, R., Ghosh, S., Dinh-Trong, T., Solberg, A.: Model-driven development using uml 2.0: Promises and pitfalls. Computer **39**(2), 59–66 (2006). doi:10.1109/MC.2006.65
5. Selic, B.: The pragmatics of model-driven development. Software IEEE **20**(5), 19–25 (2003). doi:10.1109/MS.2003.1231146
6. Apvrille, L., Courtiat, J.P., Lohr, C., de Saqui-Sannes, P.: Turtle: A real-time uml profile supported by a formal validation toolkit. Softw. Eng. IEEE Trans. 30(7), 473–487 (2004). doi:10.1109/TSE.2004.34. http://labsoc.comelec.enst.fr/turtle/ttool.html
7. Che, W., Panda, A., Chatha, K.: Compilation of stream programs for multicore processors that incorporate scratchpad memories. In: Design, automation test in Europe conference exhibition (DATE), 2010, pp. 1118–1123 (2010)
8. Chapin, J., Lum, V., Muir, S.: Experiences implementing gsm in rdl (the vanu radio description language trade;). In: Military communications conference, 2001. MILCOM 2001. Communications for network-centric operations: Creating the information force. IEEE, vol. 1, pp. 213–217 vol. 1 (2001). doi:10.1109/MILCOM.2001.985792

9. Jair Gonzalez-Pina, R.A.B.R.P.: Diplodocusdf, a domain-specific modelling language for software defined radio applications. In: Software engineering and advanced applications (SEAA), 2012 38th EUROMICRO conference on, vol. 1, pp. 213–217 vol. 1 (2012)
10. Papadopoulos, G.: Automatic code generation: a practical approach. In: Information technology interfaces, 2008. ITI 2008. 30th international conference on, pp. 861–866 (2008). doi:10.1109/ITI.2008.4588524
11. Zhu, Y., Sun, Z., Wong, W.F., Maxiaguine, A.: Using uml 2.0 for system level design of real time soc platforms for stream processing. In: Embedded and real-time computing systems and applications, 2005. Proceedings of 11th IEEE international conference pp. 154–159 (2005). doi:10.1109/RTCSA.2005.101
12. Vidal, J., de Lamotte, F., Gogniat, G., Soulard, P., Diguet, J.P.: A co-design approach for embedded system modeling and code generation with uml and marte. In: Design, automation test in Europe conference exhibition, 2009. DATE '09, pp. 226–231 (2009)
13. Yang, X.: A feasibility study of uml in the software defined radio. In: Electronic design, test and applications, 2004. DELTA 2004. Second IEEE international workshop on, pp. 157–162 (2004). doi:10.1109/DELTA.2004.10050
14. Green, P., Essa, S.: Integrating the synchronous dataflow model with uml. In: Design, automation and test in Europe conference and exhibition, 2004 Proceedings, vol. 1, pp. 736–737 Vol. 1 (2004). doi:10.1109/DATE.2004.1268954
15. Störrle, H.: Semantics of UML 2.0 activities with data-flow. In: Nordic workshop on UML (NWUML'04) (2004)
16. Dennis, J.: Data flow supercomputers. Computer **13**(11), 48–56 (1980). doi:10.1109/MC.1980.1653418
17. Kahn, G.: The semantics of a simple language for parallel programming. IFIP Cong (1974)
18. Pont, M.J.: Applying time-triggered architectures in reliable embedded systems: Challenges and solutions. e & i Elektrotechnik und Informationstechnik 125, 401–405 (2008). http://dx.doi.org/10.1007/s00502-008-0587-z
19. Hekkala, A., Harjula, I., Panaitopol, D., Rautio, T., Pacalet, R.: Cooperative spectrum sensing study using welch periodogram. In: Telecommunications (ConTEL), Proceedings of the 2011 11th international conference on, pp. 67–74 (2011)

Chapter 5
On Integrating EAST-ADL and UPPAAL for Embedded System Architecture Verification

Tahir Naseer Qureshi, De-Jiu Chen, Magnus Persson and Martin Törngren

Abstract Model-based development (MBD) is a common approach adopted in many engineering disciplines for handling complexity. For distributed microprocessor based systems MBD approaches include the use of architecture description languages (ADL's), modeling and simulation tools and tools for formal verification. To increase their combined effectiveness, the various MBD methods, tools and languages are required to be integrated with each other. This chapter addresses the connection between ADL's and formal verification in the context of automotive embedded systems. A template-based mapping scheme providing formal interpretation of EAST-ADL, an automotive specific ADL with timed automata (TA) is the main contribution providing a possibility of automated analysis of timing constraints specified for the execution behavior and events of a system. One benefit of using TA is the fact that it can also be used for generating test cases for their usage during late development phases.

5.1 Introduction

The complexity related to automotive embedded systems has increased significantly during the last few decades. While a product is composed of a large number of interconnected software and hardware components, the development process

T. N. Qureshi (✉) · D. J. Chen · M. Persson · M. Törngren
Department of Machine Design, KTH - The Royal Institute of Technology, Stockholm, Sweden
e-mail: tnqu@md.kth.se

D.-J. Chen
e-mail: chen@md.kth.se

M. Persson
e-mail: magnper@md.kth.se

M. Törngren
e-mail: martin@md.kth.se

A. Sangiovanni-Vincentelli et al. (eds.), *Embedded Systems Development*, Embedded Systems 20, DOI: 10.1007/978-1-4614-3879-3_5,

is also distributed spatially and temporally among different engineering teams. A model-based development (MBD) approach, i.e., the use of computerized models for different activities [1] is followed in various engineering domains to manage complexity. For embedded systems several powerful but disconnected MBD solutions like architectural description languages (ADL) for managing engineering information, tools like Matlab/Simulink and formal methods like model checking for various kind of analysis are used. Furthermore, a lot of faults are discovered during the late development stages (i.e., integration and testing) resulting in an increased development time and cost [2]. Inconsistency between the behavioral specifications, related constraints and requirements is one of the major factors contributing to the faults. A seamless integrated development environment incorporating several model-based methods and tools is envisioned to deal with the issues [3].

This chapter is motivated by the above mentioned needs for a seamless integrated development environment. The main objective is an architecture-centric analysis support for architectural specifications in the context of model-based development of automotive embedded systems. A method which paves a way for automated model transformation between Electronics Architecture and Software Technology- ADL (EAST-ADL); an automotive specific ADL [4] and timed automata (a formalism for verifying real-time systems) [5] is introduced. EAST-ADL provides a multi-viewed and structured information management support at multiple abstraction levels. As compared to languages like UML or AADL, EAST-ADL has a broader coverage in terms of development life-cycle. The support is both product-related, such as hardware, software and infrastructure, as well as concern related, such as requirements, safety, dependability etc. Timed automata (TA) on the other hand is a widely used formalisms for verifying real-time systems.

Provision of a generalized solution for analyzing specifications based on EAST-ADL's design level of abstraction is the main focus of the chapter. Furthermore, the work focuses on domain specific language and generic automotive systems which is in contrast with other approaches like [6, 7] focusing on either a generic language like UML or requirements organized in the form of AND/OR hierarchies.

The chapter mainly presents a mapping framework based on predefined timed automata templates. It is shown that EAST-ADL timing constraint specifications and the execution behavior of a component can be abstracted as a network of timed automata. The considerations required for the usage of the mapping scheme are also discussed. A case study of a brake-by-wire system is used to demonstrate the usage of the framework with PapyrusUML (a UML modeling tool) and UPPAAL (a TA-based model checker) as the tools for modeling and analysis respectively.

5.2 Related Work

Verification based on timed automata (TA) of automotive applications is presented in [8]. The TA models were used to derive the templates related to function execution in this chapter.

A TA-based analysis method for analyzing embedded system architectures is presented in [6]. TA models are derived from UML sequence diagrams augmented with performance data. The TA models are application specific, limited to only a few message lengths for the communication bus, etc. In addition, the generalization of the work is yet to be determined especially with the presence of *variation points* in sequence diagrams. In contrast, the work in this chapter can be applied to multiple applications focusing on EAST-ADL which does not not have variation points like UML.

A number of efforts have been carried out to enable the analysis, verification and validation of system architecture design captured in EAST-ADL. This work is an extension of [9] and [10]. In [9] the work was limited to the reaction time constraint and evaluation of the possibility of model transformation between EAST-ADL and UPPAAL. This chapter presents a few selected templates from [10] and provides an account of experiences and observations.

In [11] an effort to integrate the SPIN model checker for formal verification of EAST-ADL models is presented. The automata addressed by SPIN are untimed and therefore, less suitable for the execution behavior targeted in this chapter.

The specification and analysis of EAST-ADL models can be carried out in several ways. One way is to specify the structure using its core part and define behavior related aspects using external representations such as Simulink. This approach is used by Wang et al. [12] and Enoiu et al. [13] where timed automata is used as the external representation of EAST-ADL functional behavior. An issue with this approach is the gap between different behavioral models used during different development stages by different experts. For example, relation between timed automata models for formal verification and Simulink models for analysis and code generation. A more suitable approach is the usage of EAST-ADL native specifications for defining a common behavior and interpreting them with different formalisms used for different purposes. This approach enables the utilization of the benefits provided by EAST-ADL to a greater extent. This chapter follows the latter approach with focus on the timing model part of EAST-ADL. In [14] we have also complemented the work by addressing the internal behavior of EAST-ADL functions by transforming the EAST-ADL behavioral specifications to timed automata.

5.3 EAST-ADL and Timing Extension - Concept and Notations

EAST-ADL [4] evolved through several European projects since 2001. It complements the best industrial practices of requirements specification, system design, documentation etc. with the goal to facilitate the management of engineering information throughout a vehicle's development life cycle. The language is modular with a core part for specifying structure, and extensions for the specification of logical and execution behavior, requirements, product variations etc. This modular approach not only separates the definition of functional and non-functional aspects but also enables the use of existing tools for various development activities.

A system can be specified atfour different abstraction levels (namely vehicle, analysis, design and implementation) using EAST-ADL. For example, product line features (end-to-end functionality) and their variations are specified at the vehicle level, whereas the detailed design of functional components, connections and allocations to various hardware components is carried out at the design level. The subset of EAST-ADL addressed in this chapter is as follows.

5.3.1 EAST-ADL Core and Behavior Model

The core and behavior artifacts focused in this chapter are shown in Fig. 5.1 where the artifacts prefixed with *Behavior*:: are part of the behavior model. A concept of type and prototype is used for reusability. For example, a function representing a wheel can be instantiated as a prototype within another function representing the overall vehicle. The behavior model is mainly used to specify operational modes and triggering policy for a function or its prototype. For logical behavior (i.e., internal logics or a transfer-function between the inputs and outputs) of a function EAST-ADL relies on external representations like Simulink and SCADE.

5.3.2 Function Behavior Semantics

Every time a function is triggered it reads data at its *input* ports, *executes* computations and writes data on its *output* ports. The EAST-ADL specifications specify concurrent

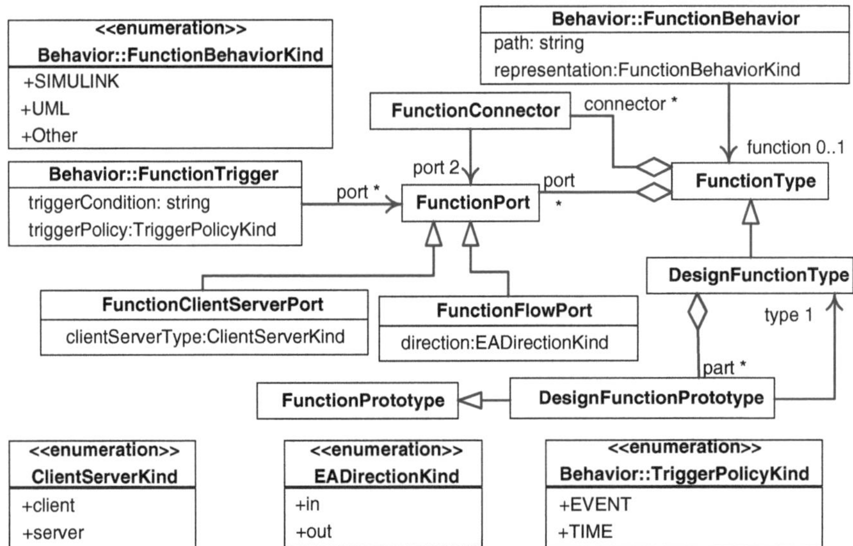

Fig. 5.1 EAST-ADL core structure and behavior model

execution and run-to-completion semantics of a function. i.e., a function runs all the steps before it starts to execute again. The behavior in terms of interruptions and preemption is considered as a part of implementation level of abstraction which is out of scope of the EAST-ADL specifications. Moreover, the ports of an EAST-ADL function have single-sized overwritable and non-consumable buffer semantics.

5.3.3 Timing Model

The timing model of EAST-ADL is derived from Time Augmented Description Language (TADL). The readers are referred to [15] for a conceptual overview of TADL. As shown in Fig. 5.2, the timing extension is based on the concepts of *event* and *event chain*. *EventFunction*, *EventFunctionFlowPort* and *EventFunction-ClientServerPort* are the three event kinds, referring to the triggering of a function by some sort of dispatcher, arrival of a data and service requested (or received) by a port. The dashed *instanceRef* relations indicate the context dependency of an artifact. For example, an EventFunctionClientServerPort is applicable on ports of prototypes instead of their respective function types. An event chain comprises of one or more *stimulus* and *response* events. An event chain can further be refined into smaller event chains called *strands* (parallel chains) or *segments* (sequenced).

Timing constraints can be specified on function executions (e.g. execution time, period, execution time budgets), event occurrences (e.g. arrival or departure of data on a function port) and event chains. The constraints addressed in this chapter are shown in Fig. 5.3. *PeriodicEventConstraint* and *DelayConstraint* in Fig. 5.3 are examples of the constraints applied on event and event chain respectively. For additional information, the readers are referred to [4].

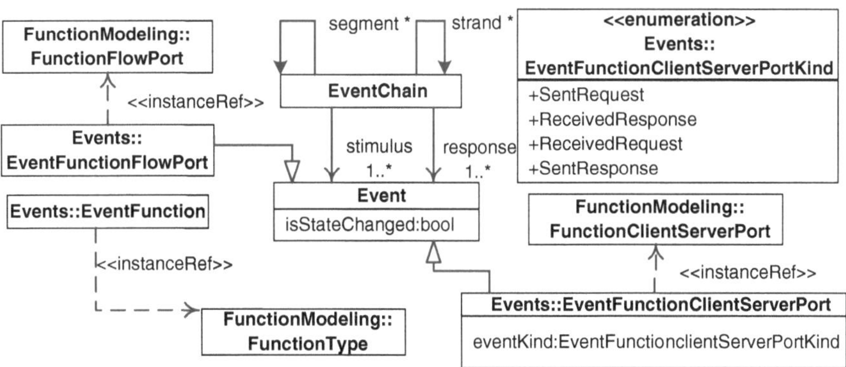

Fig. 5.2 Events and event chains in EAST-ADL

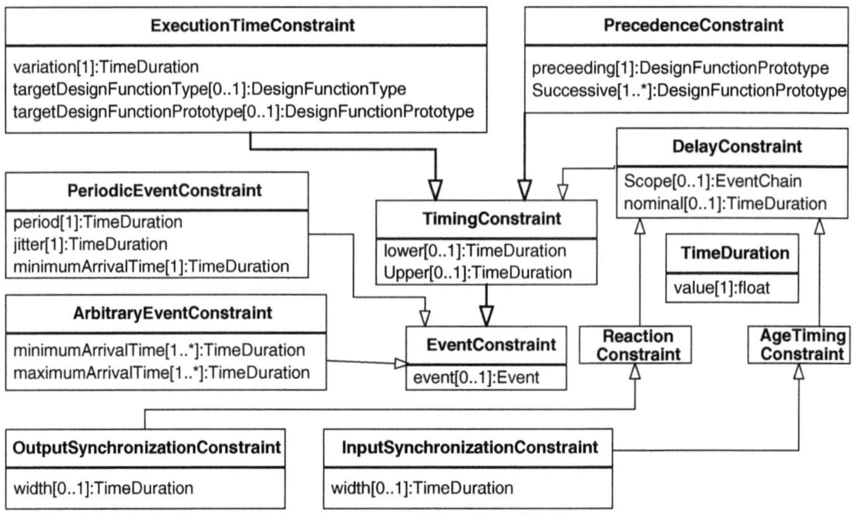

Fig. 5.3 EAST-ADL timing constraints

5.4 Timed Automata and UPPAAL

A timed automaton (TA) [5] is an automaton augmented with clocks and time semantics to enable formal analysis of real-time systems. For example, to specify the maximum time duration for which a location (or state) can remain active, a clock invariant is used. Similarly, it is also possible to specify guards based on clock values on transitions between two locations.

Often a set of TA are used in a networked form with a common set of clocks and actions. A special synchronization action denoted by an exclamation sign (!) or a question mark (?) is used for synchronization between different TA. A timed-automaton in a network is concurrent unless and until mechanisms like synchronization actions are applied. The readers are referred to [5] for a formal definition and semantics of a network of TA.

UPPAAL is a model checker based on TA for modeling, validation and verification of real-time systems. The tool has three main parts: an editor, a simulator and a verifier, for modeling, debugging and verification (covering exhaustive dynamic behavior) respectively. A system in UPPAAL is modeled as a network of TA. A subset of CTL (computation tree logic) is used as the query language in UPPAAL for verification. In addition to the generic TA, UPPAAL uses the concepts of *broadcast* channels for synchronizing more than two automata. The concepts of urgent and committed states are also introduced to force a transition without time delay. Similar to other model checking tools, UPPAAL can be used to verify (1) *Reachability* i.e., some condition can possibly be satisfied, (2) *Safety* i.e., some condition will never occur and (3) *Liveness* i.e., some condition will eventually become true. UPPAAL uses the concept of *templates* to provide reusability and prototyping of system components. Each template can be instantiated multiple times with varying parameters. The instantiation is called a *process*.

5.5 EAST-ADL and Timed Automata Relationship

Both timed automata (TA) and EAST-ADL are developed for real-time embedded systems. TA is a formalism which can be used for model-checking of real-time system. EAST-ADL has a broader coverage which includes but is not limited to structural and some behavioral aspects of embedded systems. There exist at least four different possibilities for relating EAST-ADL with timed automata. (1) Use TA for defining the behavior of a system by exploiting EAST-ADL external behavior representation support (Function Behavior in Fig. 5.1) as done in [12]. (2) Transform the EAST-ADL behavior description annex [16] to TA for a holistic behavioral analysis including logical, execution, nominal and error behavior. (3) Model timing constraints with timed automata with a suitable behavior abstraction. As the internal functional behavior and hence the associated constraints are out of scope of this work, only the timing constraints and design level of abstraction is considered, corresponding to option (3), with the following assumptions and limitations

- Only the *Functional Design Architecture* (FDA) is considered. The design level has two parts, namely Functional design architecture (FDA) and *Hardware Design Architecture* (HDA). While HDA covers hardware topology, FDA is used to model software components, middleware functions, device drivers and hardware transfer-functions. Hence, FDA together with constraints such as time budgets applied on its contained functions can provide a suitable abstraction for the target analysis.
- The sum of maximum execution time of all the functions having the same periodicity and allocated to same processing units is less then or equal to their period. This ensures that functions when refined to an implementation will be schedulable. For other cases, it is recommended to perform a schedulability analysis for each set of functions allocated on a single processor. This can be done by tools like Times (a TA-based tool for schedulability analysis). The deadline of a function execution is considered equal to its period.
- EAST-ADL supports hierarchical modeling (i.e., function types and prototypes) whereas UPPAAL does not have such support. Therefore, only one function type, i.e., the FDA (related to the software architecture) is allowed to have prototypes of other functions in its composition for the sake of simplicity.

5.5.1 Mapping Scheme

The following discusses a subset of mapping scheme between EAST-ADL and timed automata based on the experiences from a previous work [9]. An existing timed automata model of an emergency braking system (EBS) [8] was transformed to

EAST-ADL to derive the relationship followed by the validation of the mapping by transforming a brake-by-wire system, a representative industrial case study in EAST-ADL to timed automata. The fundamental concepts of the presented method is the main focus, therefore only a few templates, especially those which are related to the case-study in the next section are discussed. Therefore, the interested readers are referred to [10] for additional details.

The relationship is in the form of timed automata templates for function execution and timing constraints shown in Fig. 5.3. The templates for the timing constraints act as monitors indicating if a constraint is satisfied or not.

In the following figures, all the templates have their own clocks named 'LocalClock' and all the synchronization actions and variables like MinArrivalTime, period are their input parameters in the form of bounded integer variables. The bound represents the minimum and maximum time values in the overall system specifications. The following text covers the fundamental concepts of the presented method, however, only a few templates, especially those which are related to the case-study in the next section are discussed. Therefore, the interested readers are referred to [10] for additional details.

Event: An EAST-ADL event is modeled as a synchronization action. For example, the synchronization action *output!* for the transition from *Execute* to *Init* state in Fig. 5.4b can be considered as an event corresponding to an *EventFunctionFlowPort* referring to a port with direction out or *EventFunctionClientServerPort* with kind of either *sentRequest* or *sentResponse*. The synchronization actions correspond to a simultaneous read or write on a function's ports.

Function execution behavior: As shown in Fig. 5.4, a function can be modeled with three or two locations for time-triggered and event triggered systems respectively depending on the triggering policy of the function trigger (Fig. 5.1) specified for the function under consideration. In Fig. 5.4 , the transition from *Init* to *Execute* and *Execute* to *Finished* represent the reading and writing of data on a function port respectively. The state *Execute* abstracts the logical behavior and periodicity is modeled by the *Finished* state together with its invariant and the guard of the following transition. The parameters *maxexecTime* and *minexecTime* are obtained from *ExecutionTimeConstraint* (Fig. 5.3) where *max-* and *minexecTime* correspond to the upper and lower limits of the execution time budgets. On the other hand, the period is obtained from *PeriodicEventConstraint* applied on the *EventFunction* referring to the *DesignFunctionType* under consideration.

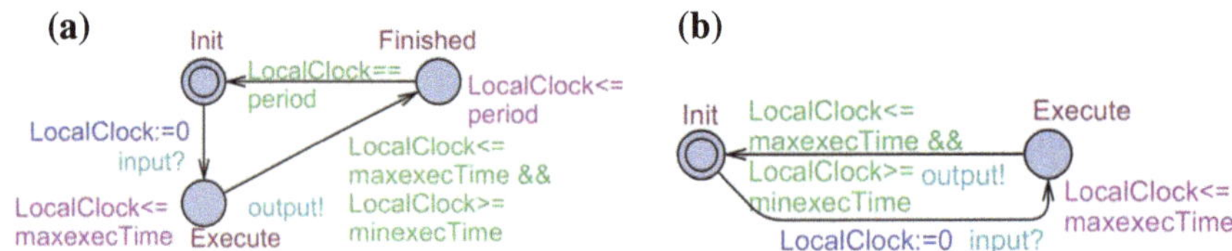

Fig. 5.4 Function templates **a** Periodic **b** Aperiodic

Timing constraints: A constraint is either satisfied or not; therefore, a minimum of four locations corresponding to initial, intermediate, success, fail states are necessary to model a constraint. On occurrence of an event, the automaton proceeds to an intermediate state. Based on the applicable guard conditions (obtained from a constraint attribute) the *fail* or *success* state is reached. The transitions to reach a *fail* or a *safe* state are enabled by clock guards and synchronization actions representing the timing bounds and event occurrences respectively. The following are two examples of the constraints applied on event and event chain respectively.

Periodic event constraint: A periodic event constraint is used to specify constraints on the periodicity of an event. An UPPAAL template for a periodic event constraint is shown in Fig. 5.5a. The three applied parameters (also shown in Fig. 5.3) are *period* (*P*), *jitter* (*J*) and the minimum arrival times of the event representing ideal, maximal and minimal time interval between occurrence of two events. The synchronization action "event?" refers to the EAST-ADL event under consideration.

Reaction constraint: A reaction constraint specifies a bound between the occurrences of stimuli and responses of an event chain. According to the EAST-ADL specifications, there exist five possible specification combinations ({upper, lower}, {upper, lower, jitter}, {upper}, {lower}, {nominal, jitter}) for the reaction constraint. The presented work considers only one combination i.e., {upper} which corresponds to the maximum time allowed between the stimulus and the response. In the reaction constraint template (Fig. 5.5b) the clock is reset when a stimulus event occurs. As soon as the response event occurs the automata transits to Fail or Success state depending on the elapsed time i.e., the 'LocalClock' value.

Rate Transition Templates: An EAST-ADL system is inherently deadlock free (Sect. 5.3.2) however, a deadlock due to blocked transitions is highly likely with the templates described earlier with functions having different execution rates. To avoid such condition, rate transition templates shown in Fig. 5.6 are introduced. The templates and the name are inspired by the Simulink[1] rate-transition block.

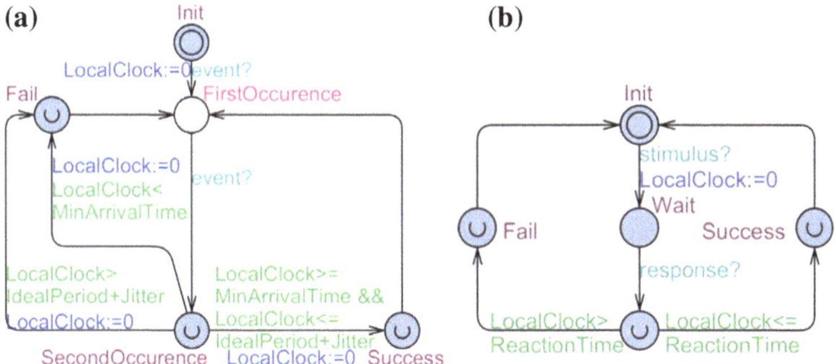

Fig. 5.5 Periodic event and reaction constraint templates

[1] http://www.mathworks.se/help/toolbox/simulink/slref/ratetransition.html

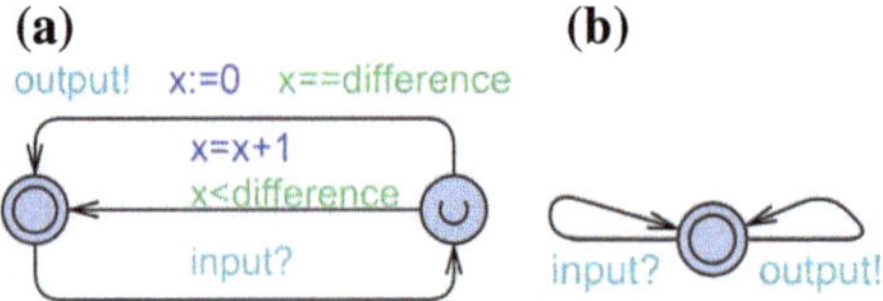

Fig. 5.6 Rate transition templates **a** Fast to slow rate transition **b** Slow to fast rate transition

The template in Fig. 5.6a is used when the sender is running at a faster rate (less period) than the receiver. The actions *input?* and *output!* correspond to the input from the sender and output to the receiver respectively. The *difference* parameter is the *difference of frequencies* obtained by dividing the period of the receiver with that of sender. For the case where the sender has low frequency, the template in Fig. 5.6b is used.

5.5.2 Usage and Automation Considerations

The following is required for the transformation and its automation:

1. A pre-defined set of templates in UPPAAL with different combination of channel types (i.e., 'chan' and 'broadcast chan' in UPPAAL) and inclusion/exclusion of synchronization actions from function temples. In this way the automation process will simply require template instantiations as UPPAAL process corresponding to EAST-ADL functions. For example, for a function in EAST-ADL representing a sensor having no input, an UPPAAL function template without 'input?' synchronization will be used. Similarly, if a constraint is applied on an EAST-ADL event or if it is transmitted to multiple receivers then the corresponding channel type in UPPAAL should be 'broadcast' type.
2. An appropriate rate transition template should be used for two functions executing at different rates and communicating with each other. This in turn requires declaration of additional channels. For example, consider a process 'A' in UPPAAL is communicating with process 'B' using channel 'channel1'. If a rate transition 'RT1' is used then an additional channel, i.e., 'channel2' is defined. In this case 'channel1' is used between 'A' and 'RT1' whereas 'channel2' is used between 'RT1' and 'B'. If the two communicating functions have the same frequency then no rate-transition process is required.
3. The minimum resolution of time has to be decided beforehand where time values are declared as a multiple of minimum time unit. For example, if the minimum unit is defined as 1 μs then 1 ms will be written as 1,000.

5.5.3 System Verification

With the introduced relationship between EAST-ADL and timed automata, the analysis becomes verification of a safety property in the following form: (1) A given constraint is satisfied *iff* for all initial conditions, the state "Fail" is never reached for all cases. (2) A system is free of any inconsistencies *iff* there is no deadlock and all the constraints are satisfied. ToS verify a given set of timing constraints of a system specification, the following two query language syntaxes have to be used.

- *A [] (not deadlock)* to verify if there exist any deadlock. Three possible reasons for a deadlock can be (1) the absence of one or more rate-transition templates in case of different frequency between two communicating functions and (2) an incorrect synchronization channel type and (3) Incompleteness of the specifications being verified.
- *A [] (not XX.Fail)* to verify that a timing constraint modeled with an UPPAAL process named XX never reaches the failed state. In case a fail state is reached, the timing constraints have one or more inconsistencies.

5.6 Brake-by-Wire Case Study

The brake-by-wire (BBW) system is a representative industrial case study. This case has been used in several EAST-ADL related projects funded by the European Commission. It provides coverage of EAST-ADL artifacts and methodology at multiple abstraction levels. A simplified version of the case study is shown in Fig. 5.7 which is a snapshot from its EAST-ADL (UML profile) implementation. For simplicity, three actuators, ABS functions and their connections are not shown in the figure.

In the simplified system, the Brake Torque Calculator (BTC) periodically (period = 10 ms) calculates the desired torque utilizing the percent value of the brake pedal. This

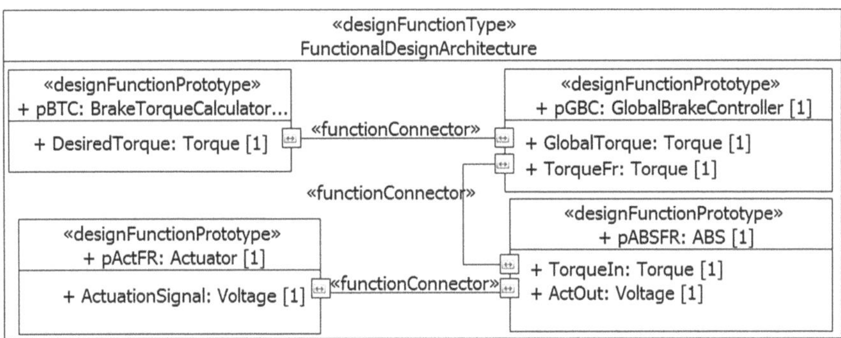

Fig. 5.7 UML implementation of the brake-by-wire system

desired torque is utilized by the GlobalBrakeController (GBC) which periodically (period = 20 ms) calculates the torque for each wheel. The ABS functions on each wheel are responsible to control the locking of the wheel and provide the required braking torque. Both ABS and the actuators are triggered on the events related to the arrival of a data on their ports. This implies that after transformation to UPPAAL BTC and GBC will be represented by the template shown in Fig. 5.4a and the ABS by the one in Fig. 5.4b. In addition a rate transition template will also be required between BTC and GBC.

Four different events listed in Fig. 5.8 are defined for the EAST-ADL model. Furthermore, only one event chain (see also Fig. 5.2) is defined with the events 'CalculatedTorque' and 'actuation1' as the stimulus and response respectively. The constraints are listed in Fig. 5.9.

An UPPAAL model of a subset of the constraints and functions described above is shown in Fig. 5.10 where the red color represents the active locations during a simulation . The model is manually generated. Compare the process names, synchronization actions and constraints, i.e., PCC1, RC1 and PEC1 with the discussion and figures earlier in this section.

Fig. 5.10 shows three constraints i.e., PEC1 (Periodic Event Constraint), PCC1 (Precedence Constraint) and RC1 (Reaction Constraint). To verify the consistency the CTL syntax as described in Section 5.5.3 were used. For example, *A[] (not deadlock)* to check if the system is deadlock free and *A[] (not PEC1.Fail)* to check if

Event Name	Type	Attributes
BTCTriggerEvent	EventFunction	TargetFunctionPrototype=pBTC
GBCTriggerEvent	EventFunction	TargetFunctionPrototype=pGBC
CalculatedTorque	EventFunctionFlowPort	TargetFunctionFlowPort=DesiredTorque TargetFunctionPrototype=pBTC
actuation1	EventFunctionFlowPort	TargetFunctionFlowPort=ActOut TargetFunctionPrototype=pABSFR

Fig. 5.8 Events

Constraint Name	Type	Attributes
BTCExecution	Execution time	TargetFunctionPrototype=pBTC , Lower = 3 , Upper =5
GBCExecution	Execution time	TargetFunctionPrototype=pGBC, Lower = 2, Upper =6
ABSFRExecution	Execution time	TargetFunctionPrototype=pABSFR, Lower = 2, Upper =3
ABSFLExecution	Execution time	TargetFunctionPrototype=pABSFL, Lower = 2, Upper =3
ABSRRExecution	Execution time	TargetFunctionPrototype=pABSRR, Lower = 2, Upper =3
ABSRLExecution	Execution time	TargetFunctionPrototype=pABSRL, Lower = 2, Upper =3
RC1	Reaction	Scope = EC1, ReactionTime = 50 ms
PEC1	Periodic event	TargetEvent= CaclulatedTorque, MinArrivalTime= 3ms, IdealPeriod = 10 ms, Jitter = 9 ms
PCC1	Precedence	Preceding = CalculatedTorque, Successive = actuation1

Fig. 5.9 Timing and event constraints

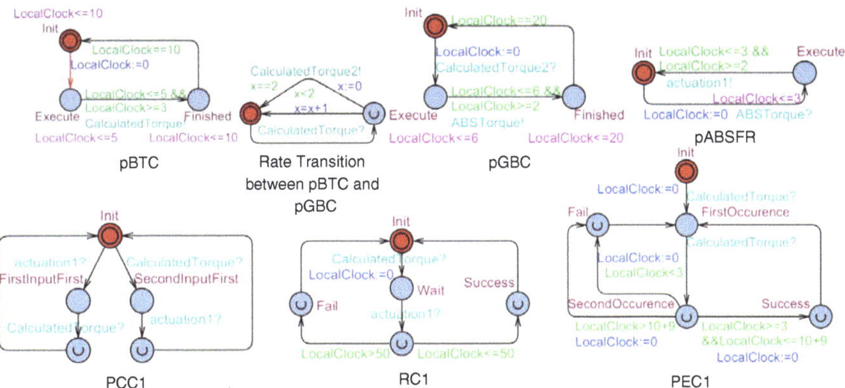

Fig. 5.10 Brake-by-wire model in UPPAAL

the periodic event constraint is satisfied. All the properties were found to be satisfied using the UPPAAL verifier. The time taken for the verification was less than a second.

5.7 Discussion

A method to analyze consistency of timing constraints specified using EAST-ADL is presented. The proposed mapping scheme is a basis for automated transformations between EAST-ADL and timed automata based tools. We have experimented with different model transformation technologies such as Atlas Transformation Language (ATL) or Model Query Language (MQL) [9]. Transformation of EAST-ADL specifications is a complex and non-trivial task due to the distribution of information among its extensions. One of the main benefits of the presented approach is that it reduces the effort related to model-transformation and related complexity. Instead of creating a new TA process for every timing constraint, the transformation requires only instantiation of the templates. This is especially true for imperative transformation languages like MERL (MetaEdit+ Reporting Language). Furthermore, the template-based approach is also a scalable solution where the templates can be extended for as many events as required to be considered for an event chain. The extension mechanisms are discussed in [10].

While there exist otherapproaches targeting different types of constraints for example, dependency graph for checking function precedence [17], the presented work is suitable in the context of MBD. The use of TA can enable bridging the gap between the design and testing phase. Generation of test cases from the timed automata analysis of EAST-ADL specifications for carrying out testing at different design phases is one of the future possible extensions.

At first glance, the use of techniques like scheduling analysis might seem to be more appropriate than the use of timed automata. The use of such techniques are applicable at the implementation level of EAST-ADL which in turn corresponds to the AUTOSAR standard [18] instead of the design level of abstraction addressed by this chapter. To enable analysis such as schedulability, the design level of abstraction has be augmented with additional details as presented in [19]. In addition, the concurrency of functions at the design level of abstraction and the fact that there exists an *n-to-m* relationship between EAST-ADL functions and AUTOSAR software components as well as runnables [20, 21] further motivates the use of timed automata based analysis.

In Sect. 5.5, three different possibilities for relating EAST-ADL and timed automata were mentioned. Based on the experiences from the work presented in this chapter, the possibility of transforming the EAST-ADL behavior description annex (BDA) to timed automata has also been studied [14]. The BDA provides support for relating different kind of behavior such as execution behavior and timing constraints considered in this chapter, logical behavior, and error behavior specified by the EAST-ADL's dependability modeling extension. The templates described in this chapter can serve as a monitor to verify different types of timed behavioral.

The main focus of the work presented in this chapter was on exploring a relationship between EAST-ADL and timed automata as a step towards an integrated environment providing architecture-centric development of automotive embedded systems in a seamless manner. This also implies that aspects like state space coverage were not the major focus however, discussed in [14]. The timed automata semantics and the assumptions presented in this chapter pose some limitations on EAST-ADL modeling and transformation work. For example, the function templates assumes one input and one output port which in turn requires that a function can only be triggered by arrival of data on one port only etc. in contrast to EAST-ADL support for multiple triggers. Another issue related to the fast to slow rate transition template is that it is only applicable for the cases where the frequencies are multiples of each other. Similarly, aspects like jitter are considered to a limited extent. Efforts to overcome this limitations are one of the planned future activities.

Other issueswhich require further investigation are automatic test-case generation directly from EAST-ADL specifications, combined analysis of internal and execution behavior and methods for transferring the analysis results back for storing them as part of the EAST-ADL model using its verification and validation extension.

References

1. Törngren, M., Chen, D., Malvius, D., Axelsson, J.: Automotive embedded systems handbook. In: Model-Based Development of Automotive Embedded Systems. CRC Press, Boca Raton, FL (2009)
2. Lönn, H., Freund, U.: Automotive embedded systems handbook. In: Automotive Architecture Description Languages, CRC Press, Boca Raton, FL (2009)
3. Broy, M., Feilkas, M., Herrmannsdoerfer, M., Merenda, S., Ratiu, D.: Seamless model-based development: from isolated tools to integrated model engineering environments. Proc. IEEE **98**(4), 526–545 (2010)

4. The ATESST2 Consortium: EAST ADL 2.0 Specification. Project Deliverable D4.1.1. http://www.east-adl.info/repository/EAST-ADL2.1/EAST-ADL-Specification_2010-06-30.pdf (2010)
5. Bengtsson, J., Yi, W.: Timed automata: semantics, algorithms and tools. In: Lectures on Concurrency and Petri Nets, LNCS, vol. 3098, pp. 87–124. Springer, Berlin (2004)
6. Hendriks, M., Verhoef, M.: Timed automata based analysis of embedded system architectures. In: Parallel and Distributed Processing Symposium, 2006. IPDPS 2006. 20th, International, p. 8 (2006)
7. Ponsard, C., Massonet, P., Molderez, J.F., Rifaut, A., Lamsweerde, A.V., Van, H.T.: Early verification and validation of mission critical systems. Form. Methods Syst. Des. **30**(3), 233–247 (2007)
8. Montag, P., Nowotka, D., Levi, P.: Verification in the design process of large real-time systems: a case study. In: Automotive Safety and Security 2006, Stuttgart (Germany), October 12–13, 2006, pp. 1–13 (2006)
9. Qureshi, T.N., Chen, D.J., Persson, M., Törngren, M.: Towards the integration of UPPAAL for formal verification of EAST-ADL timing constraint specification. In: Workshop on Time Analysis and Model-Based Design, from Functional Models to Distributed Deployments (2011)
10. Qureshi, T.N., Chen, D.J., Törngren, M.: A timed automata-based method to analyze EAST-ADL timing constraint specifications. In: Modelling Foundations and Applications, Lecture Notes in Computer Science, vol. 7349, pp. 303–318. Springer, Berlin (2012)
11. Feng, L., Chen, D.J., Lönn, H., Törngren, M.: Verifying system behaviors in EAST-ADL2 with the SPIN model checker. In: Mechatronics and Automation (ICMA), 2010 International Conference on, pp. 144–149 (2010)
12. Kang, E.Y., Schobbens, P.Y., Pettersson, P.: Verifying functional behaviors of automotive products in EAST-ADL2 Using UPPAAL-PORT. In: Computer Safety, Reliability, and Security, LNCS, vol. 6894, pp. 243–256. Springer, Berlin (2011)
13. Enoiu, E.P., Marinescu, R., Seceleanu, C., Pettersson, P.: ViTAL : A verification tool for EAST-ADL models using UPPAAL PORT. In: Proceedings of the 17th IEEE International Conference on Engineering of Complex Computer Systems. IEEE Computer Society Press (2012)
14. Qureshi, T.N.: Enhancing model-based development of embedded systems: modeling, simulation and model-transformation in an automotive context. Trita-mmk, issn 1400–1179; 2012:16, isbn 978-91-7501-465-4, Department of Machine Design, KTH - The Royal Institute of Technology, Sweden (2012)
15. Stappert, F., Jonsson, J., Mottok, J., Johansson, R.: A design framework for End-To-End timing constrained automotive applications. Embedded Real-Time Software and Systems (ERTS'10) (2010)
16. The MAENAD consortium: language concepts supporting engineering scenarios. Project Deliverable D3.1.1. http://www.maenad.eu/publications.html (2012)
17. Zhao, Y.: On the design of concurrent, distributed real-time systems. Ph.D. thesis, EECS Department, University of California, Berkeley (2009)
18. AUTOSAR Consortium: AUTomotive Open System ARchitecture. http://www.autosar.org
19. Anssi, S., Tucci-Pergiovanni, S., Mraidha, C., Albinet, A., Terrier, F., Gerard, S.: Completing EAST-ADL2 with MARTE for Enabling Scheduling Analysis for Automotive Applications. Conference ERTS, In (2010)
20. Cuenot, P., Frey, P., Johansson, R., Lönn, H., Reiser, M.O., Servat, D., Tavakoli Kolagari, R., Chen, D.: Developing automotive products using the EAST-ADL2, an AUTOSAR compliant architecture description language. In: Proceedings of the 4th European Congress ERTS (Embedded Real Time Software) (2008)
21. Qureshi, T.N., Chen, D., Lönn, H., Törngren, M.: From EAST-ADL to AUTOSAR. Tech. Rep. TRITA-MMK 2011:12, ISSN 1400–1179, ISRN/KTH/MMK/R-11/12-SE, Mechatronics Lab, Department of Machine Design, KTH, Stockholm, Sweden (2011)

Chapter 6
Schedulability Analysis at Early Design Stages with MARTE

Chokri Mraidha, Sara Tucci-Piergiovanni and Sebastien Gerard

Abstract The construction of a design model is a critical phase in real-time systems (RTS) development as the choices made have a direct impact on timing aspects. In traditional model-based approaches, the design relies largely on the designer experience. Once the design model is constructed, a convenient schedulability test has to be found in order to ensure that the design allows the respect of the timing constraints. This late analysis does not guarantee the existence of a test for the given design and does not allow early detection of unfeasible designs. In order to overcome this problem, this chapter proposes the first UML/MARTE methodology for schedulability-aware real-time software design models construction.

6.1 Introduction

Model-based approaches for real-time systems (RTS) development aim at going from requirements specification to binary code production with the insurance of respecting the functional and non-functional requirements of the system. These model-based approaches (e.g. [1–3]) introduce a number of intermediate models between the requirements and the binary code. Requirements are usually formalized with use case scenarios. Even when modeled with other formalisms, critical scenarios that represent the system response to external stimuli are specified along with the system response deadlines. The functional model aims at representing functional blocks and

C. Mraidha (✉) · S. Tucci-Piergiovanni · S. Gerard
CEA, LIST, 91191 Gif-sur-Yvette CEDEX, France
e-mail: chokri.mraidha@cea.fr

S. Tucci-Piergiovanni
e-mail: sara.tucci@cea.fr

S. Gerard
e-mail: sebastien.gerard@cea.fr

A. Sangiovanni-Vincentelli et al. (eds.), *Embedded Systems Development*,
Embedded Systems 20, DOI: 10.1007/978-1-4614-3879-3_6,

their interactions and to show how functional blocks participate in above defined scenarios.

The functional model is then refined in a design model that introduces the mechanisms/patterns for the realization of the functional model on the underlying platform. Actually, the design model defines the architecture of the system still being independent of specific platforms. To this end abstracted resources and services are assumed. Threads, which are the unit of execution handled by OS platform services, have to be structured to determine the threading strategy of the system. By defining a mapping of functional blocks on threads, it is defined how the system reacts to external stimuli. Finding a convenient threading strategy is a complex task. The designer has a set of real-time design patterns [4] that he can use to determine the adequate number of threads and the good grouping of functions to threads towards the definition of a of the system. This concurrency model is of primary importance with respect to the system's . In order to ensure that the concurrency model satisfies timing requirements, a timing validation , like schedulability analysis , is necessary at this point. However, this timing validation can lead to the following problems: (1) the resulting design model can be too difficult or even impossible to analyze [5], (2) the resulting design model is analyzable but the designer would like to explore other possible design models, to explore several candidates from a schedulability point of view. From these considerations, it follows that it is necessary to guide the designer in the construction of an analyzable design model, i.e., a design model for which a schedulability test there exists and possibly support the cohabitation of several analyzable design candidates in order to support a comparative analysis.

This chapter focuses on a real-time methodology, called Optimum, offering a UML front-end for the designer that conforms to a formal model for schedulability analysis and supports the evaluation of different architecture candidates. In order to offer such a UML front-end, standard UML needs, on one hand, to be extended to schedulability concepts and, on the other hand, to be restricted in the use of some elements to express a precise semantics. To this purpose Optimum uses MARTE [6], a standard specialization of UML for real-time and embedded systems development. It provides support for specification, design and verification/validation stages. Actually, before MARTE, the UML SPT profile [7] was available. In [8] authors show how SPT was not sufficient to express some basic concepts for schedulability and they propose to extend the profile in ad hoc way with those concepts. Optimum, which uses similar concepts used in [8], does not need to extend MARTE. MARTE proved to be enough rich to express all the concepts needed to build schedulability models. However, as the MARTE profile is quite rich, a restriction of the language is necessary in order to delimit the usage of MARTE concepts in the context of the Optimum methodology. Nevertheless, this restriction of the language has to preserve the compatibility with MARTE standard. In the best of our knowledge, Optimum is the first methodology for schedulability analysis using a standard specialization of UML.

The chapter is organized as follows. Section 6.2 presents an overview of the Optimum methodology. Section 6.3 presents the formal schedulability analysis model considered in the Optimum methodology. In Sect. 6.4, a detailed description of the methodology and its conformance to the formal schedulability model is given.

Section 6.5 illustrates the methodology usage on an automotive example. Related work is discussed in Sect. 6.6. Section. 6.7 concludes the chapter.

6.2 Overview of the Optimum Process

The Optimum process guides the selection and the design of the concurrency model on the base of hard real-time constraints, expressed at the end of the requirement specification phase, towards the construction of concurrency models analyzable in automated way. To this end the methodology defines a process depicted on Fig. 6.1 for the generation of the architecture model specifying a concurrency model and a deployment model on the execution platform .

The methodology process has as entry point two artifacts: (1) the functional model, i.e., a description of system end-to-end scenarios and (2) the description of timing requirements . Both artifacts are assumed to be available at this stage of the development process. Some basic characteristics of the hardware abstraction layer are also assumed to be available, as the type of the execution hosts that constitute the resource platform for the execution of identified scenarios. In fact, the type of execution host is used to allow the assignment of time budgets to the functions in the scenarios. This available information is organized in a specific UML model enriched with the MARTE profile called Workload model . The Workload model specifies

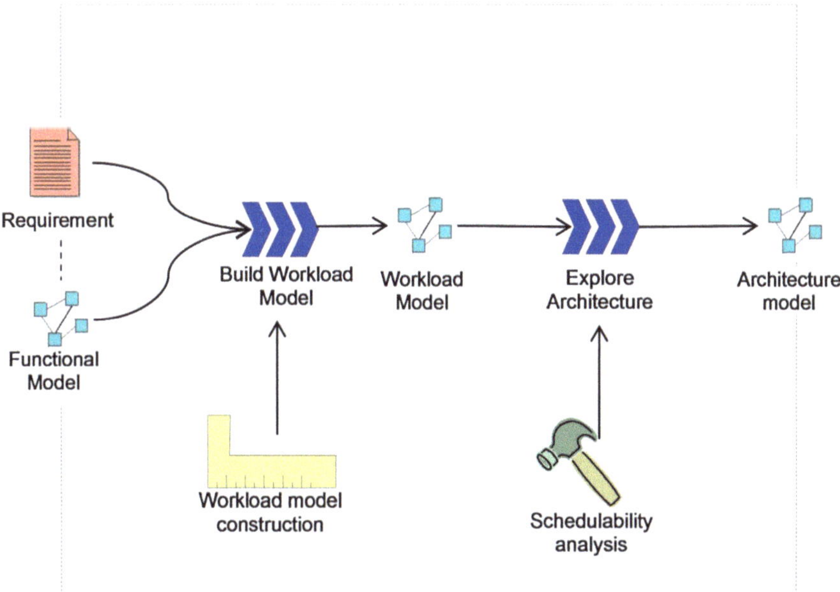

Fig. 6.1 Optimum process overview

two concerns: a graph of function activations along with execution time budgets and end-to-end deadlines, and an abstract layer representing the execution platform. In order to obtain a satisfactory architecture model, the methodology provides at this point a way of exploring different alternatives by setting a so-called analysis context. Each analysis context contains an architecture model subject to evaluation. The architecture model contains two views. The first view is generated by transformation of the graph of function activations contained in the Workload model. This transformation, also called, defines the tasks and the functions allocated on them. The second view represents the software platform resources in terms of tasks along with their scheduling parameters , and the hardware resources in terms of processing and communication. More in detail, this architecture model describes the evaluated concurrency model by schedulability analysis techniques and embeds also the following information (1) the mapping of functional blocks on tasks, (2) the level of threading for each functional block and possible synchronization resources needed to manage multithreading (if any), (3) the needed OS support in terms of the scheduler algorithm and preemption capabilities, (4) the policy for serving external events, (5) the needed to protect . The evaluation of the dynamics paradigm can cover different aspects such as resource utilization, response times, jitters, etc.

6.3 The Schedulability Model

In this section we present the formal model to analyze the schedulability of a real-time system. We introduce the formal notation to later show the conformance of the UML/MARTE model to the presented formal model.

The presented model captures a distributed system with fixed-priority tasks and task dependencies. This model is the one assumed by the test of Palencia and Harbour [9] and available in the MAST open source tool. The model is quite general, considering that fixed-priority scheduling is the most common scheduling algorithm available in practice. Let us also remark that whenever the tasks set under consideration is deployed on a single CPU, the Palencia and Harbour test is anyway applicable, reducing to the Lehoczky test [10]. Tasks are characterized by an arrival pattern that can be periodic or aperiodic with minimum inter-arrival time. We consider an event-activation paradigm. Given a set of external events, each external event triggers one computation, the system response for that event. One single task may execute the whole response or the response may be "segmented" by a sequence of tasks possibly spanning several computation resources. In case of segmentation, a number of tasks are not directly triggered by the external event. For those tasks, the event-activation paradigm states that the trigger for activation is execution's end of the previous task. In case more than one trigger there exists for the same task, AND/OR semantics may be specified. The OR semantics implies that the task is triggered by the execution's end of any of the previous tasks, while the AND semantics implies that the task is triggered only when all the previous tasks have completed. In this chapter we restrict our attention to the OR semantics. The reason of this choice is twofold: on one hand this assumption let simplify the model and makes the presentation clearer, on the other hand the test of Palencia and Harbour does not fully support the AND semantics (the AND semantics can be specified only among events of the same system response, events with the same rate), therefore we could not use available open source tools to analyze a model containing AND semantics.

The model is characterized by a set of events $E = (e_1, e_2, \ldots, e_n)$, a set of tasks $T = (\tau_1, \tau_2, \ldots, \tau_m)$, a set of computational/communication resources $R = (r_1, r_1, \ldots, r_l)$, a characterization of event occurrences and a characterization of task executions. The characterization of event occurrences is constituted by a set of tuples, one for each event $e_i \in E$ of the form $(e_i, t_i, R_i, D_i, WR_i)$, where e_i is the event whose occurrence being characterized, t_i is the period of the event, R_i is the response to the event e_i. R_i is a total order of tasks $(T_i, \rightarrow_i)$, where $\emptyset \subset T_i \subseteq T$ is the set of tasks to be executed in response to the event e_i. For any two tasks $\tau_h, \tau_k \in T$: the task τ_h has to be executed before τ_k, in the response R_i if and only if $\tau_h \rightarrow_i \tau_k$. D_i is the end-to-end deadline for response R_i. WR_i is the worst case response time for response R_i.

Characterization of task executions is constituted by a set of tuples, one for each task $\tau_h \in T$ of the form $(\tau_h, P_h, r, CS_h, C_h, B_h, WTR_h)$, where: τ_h is the task being characterized, P_h is the assigned priority, $r \in P$, is the computational/communication resource the task is executed on, C_h is the computational cost of the task (not considering any contention time), CS_h: ordered list of critical sections $(cs_{h1}, cs_{h2}, ..., cs_{hl})$ the task will pass through. A critical section represents an interval of non-preemption. The scheduler cannot interrupt a task in critical section even if a higher priority task is ready for execution. Critical section cs_{hi} has a duration $c_{hi} \leq C_h$. B_h is the blocking time, i.e., the time a task is blocked by another task with lower priority. This happens during concurrent access to critical sections: when a lower priority task is in critical section as it cannot be preempted until it releases its lock. A higher priority task has to wait the lock release before acquiring the lock. WTR_h is the worst-case response time of the task τ_h.

Note that in our model the same task can appear in more than one response and that each response establishes a total order in the tasks execution during the response. For the model to be valid, the following property must hold:

Partial-Order Property: the set of event characterizations induces a cyclic-free partial order on the set of tasks T. This property means that a unique order between dependent tasks can be established.

Schedulability condition: The model is schedulable if and only if all responses R_i have a worst-case response time $WR_i \leq D_i$.

In our model the following additional assumptions hold:

Assumption 1. (Fixed-priority scheduler). We assume that for each computational/communication resource there exists one scheduler arbitrating the access to the resource by a fixed-priority policy.

Assumption 2. (Priority-ceiling). We assume that passive resources, accessed in critical section regions, are protected by a priority ceiling protocol (to avoid priority inversion).

Assumption 3. (CAN-like channels). We assume that communication channels are arbitrated by a CAN-like protocol, where messages inherit the priority level of sending tasks. At destination buffered messages are dequeued on priority basis.

6.4 Detailed Optimum Methodology

In this section a description of the standard modeling language on which the Optimum methodology is based is given. Then the models used and produced by the Optimum process will be characterized. A third subsection will present the conformance of the produced software architecture model with the formal schedulability analysis model presented in Sect. 6.2. At last, the software architecture exploration phase will be detailed.

6.4.1 Modeling Language Description

The MARTE standard [6] provides all the concepts needed for the analysis and design of real-time and embedded systems (RTES) in a model-based approach. As a standard, MARTE proposes 158 concepts in order to cover a large broad of development needs for RTES. Furthermore, MARTE gives several concepts with a close semantics to represent a common global notion. For instance, a schedulable resource could be a SwSchedulableResource from MARTE::SRM or a SchedulableResource from MARTE::GRM. Differentiation of these nearby equivalent concepts can be difficult to make. Thus, a MARTE-based methodology should specify a subset of MARTE concepts that is sufficient for its purpose.

This section introduces the concepts of MARTE on which the Optimum methodology relies. The usage of these concepts in the methodology is then restricted by a profile, called the MARTE4Optimum profile.

6.4.1.1 The MARTE4Optimum Profile

Table 6.1 enumerates the 14 useful concepts for the Optimum methodology out of the 158 ones offered by MARTE. The needed concepts deal with platform re-sources modeling, schedulability analysis modeling and allocation modeling. MARTE stereotypes extend too general metaclasses of UML. Allowing such a large applicability may make validation of methodological rules more complex and limit automation of a refinement process for the methodology. The Fig. 6.2 below represents a stereotype specialization principle. A MARTE4Optimum GaWorkloadBehavior stereotype that specializes the MARTE GaWorkloadBehavior stereotype is created. The specialization does not add any property and preserves all properties of the original stereotype but the extension of the NamedElement metaclass. We

Table 6.1 Optimum restriction of MARTE subset

MARTE4Optimum stereotype	UML extensions
Alloc::Allocate	Abstraction
Alloc::Allocated	CallAction, Property
GRM::SchedulableResource	Property
GQAM::GaPlatformResources	Class
GQAM::GaWorkloadBehavior	Activity
GQAM::GaWorkloadEvent	AcceptEventAction
HRM::HwComputing::HwProcessor	Class
SAM::SaAnalysisContext	Package
SAM:: SaEndToEndFlow	ActivityPartition
SAM:: SaCommHost	Connector
SAM::SaExecHost	Property
SAM::SaSharedResource	Property
SAM::SaCommStep	ControlFlow
SAM::SaStep CallBehavior	Action

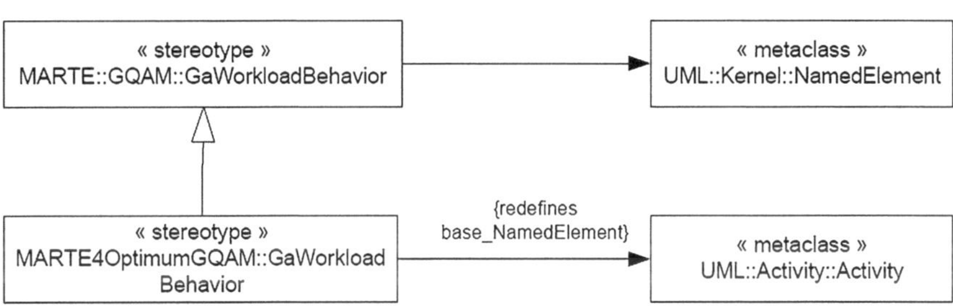

Fig. 6.2 Stereotype specialization principle

make use of the UML redefinition capability to redefine this extension with a more special one (a UML Activity is also a UML NamedElement). Therefore, we restrict the application of the GaWorkloadBehavior stereotype to a UML Activity which is a rule of the Optimum methodology described in the following section.

Table 6.1 lists the UML extensions specialization for the MARTE subset.

6.4.1.2 MARTE Compatibility

The MARTE4Optimum profile is constructed with respect to the following rules: (1) MARTE4Optimum stereotypes and general MARTE stereotypes have the same name, (2) MARTE4Optimum stereotypes inherit from MARTE stereo-type properties and no additional property is added to them, (3) UML metaclass used in the extension redefinition is necessary a specialization of the UML meta-class used in the general MARTE stereotype extension.

Compatibility with the MARTE profile is then preserved by the MARTE4Optimum profile. The export of a MARTE4Optimum model in MARTE can be easily realized by a one-to-one mapping consisting in unapplication of Optimum4MARTE stereotypes and the application of the corresponding MARTE stereotypes.

6.4.2 The Optimum Models

6.4.2.1 The Workload Model

The Workload model is constituted of a workload behavior specifying the con-trolled sequence of actions triggered by external stimuli enriched with timing in-formation. The construction of the workload behavior proceeds by the generation of a UML Activity diagram containing a canonical form of the controlled sequence of actions contained in the functional model. For the canonical form, the following properties hold.

Properties on the Activity diagram are:

P1—The subset of activity diagram elements used is: AcceptEventAction nodes, CallBehaviorAction nodes, FlowFinalNode, MergeNode, ControlFlow, ActivityPartition. All other activity diagram elements are not used.

P2—Events are modeled as UML AcceptEventActions that have UML Triggers referencing the Events. These events are modeled with UML SignalEvents.

P3—In response to an event, the invocation of a function (modeled as UML Activity) is modeled with UML CallBehaviorAction. The CallBehaviorAction represents the call to the UML Activity representing the function.

P4—The last action to be executed in response to an event is always followed by exactly one UML FinalFlowNode.

P5—For each path connecting an AcceptEventAction to a FlowFinalNode, there exist an ActivityPartition containing all the nodes of the path.

P6—An AcceptEventAction has exactly one outgoing control flow that targets a CallBehaviorAction.

P7—Each path of the Activity begins with an AcceptEventAction node, terminates with a FlowFinalNode.

P8—The controlled sequence of actions does not contain cycles.

Properties on MARTE-based annotations are:

P9—The Activity is stereotyped «GaWorkloadBehavior».

P10—Each UML AcceptEventAction is stereotyped «GaWorkloadEvent». The arrival pattern (periodic, sporadic) of the event is specified in the arrivalPattern property. For the periodic pattern the period is specified following the Value Specification Language (VSL) syntax periodic=(value, unit) where value is a nu-merical value and unit is the time unit used (e.g. s, ms,..). Similarly, the minimum inter-arrival time is specified with the following syntax sporadic(value, unit).

P11—Each UML ActivityPartition is stereotyped «SaEndToEndFlow». An end-to-end deadline is specified in the end2EndD property of the «SaEndToEndFlow» stereotype. This property is a duration specified with the following VSL syntax (value, unit).

P12—Each action of type CallBehaviorAction is stereotyped «SaStep». Each step has a computational budget (execTime property) specified for a given type of execution host (host property).

At this stage of the process, the platform is not determined yet because no deployment is specified. However, as previously explained we need to specify an estimation of computation time budget for the actions stereotyped «SaStep». A computation time budget makes sense if it is related to a type of execution host. This type of execution host is modeled with a UML Class stereotyped «HwProcessor». We can notice that this type can be modeled in model library and reused in several workload models. This type of execution host will be used in the architecture model for the definition of the platform processing resources.

6.4.2.2 The Architecture Model

The architecture model enriches the workload model with (1) an explicit con-currency strategy and (2) a deployment model on a target hardware platform.

The concurrency model

A new UML Activity is built for the modeling of the concurrency strategy. This activity contains the information provided in the workload model and adds to it the information on the mapping of function executions (steps) to tasks. Moreover, this UML activity will explicitly describe synchronization resources used to treat contention in the multi-threading case. All the properties defined above for the characterization of the UML Activity for the workload model specification are valid for this new Activity. Yet, the following additional properties characterize the new UML Activity for the concurrency strategy modeling:

P13—A UML ActivityPartition is created for each task. The property "represents" of this ActivityPartition is valuated with a reference to the corresponding element modeling the task in the platform resources (see next section).

P14—Each CallBehaviorAction is mapped to at least one task. This mapping is modeled by including the CallBehaviorAction in the ActivityPartitions representing the tasks.

P15—Let us consider two ActivityPartitions representing tasks t1 and t2 and two sets of CallBehaviorActions A1 and A2. If there exists a path from one or more actions of A1 to one or more actions of A2 and there exists a path from one or more actions of A2 to one or more actions of A1, then the mapping actions to tasks is not valid.

Additional properties on MARTE-based annotations are:

P16—A UML Package is stereotyped «SaAnalysisContext»is created. The platform property of this stereotype is valuated with a reference to the classifier stereotyped «GaResourcesPlatform»that models the platform resources (see next section). The workload property is valuated with a reference to the Activity representing the concurrency model. Note that this stereotype is the key language concept to support architecture exploration as detailed below.

P17—Each ActivityPartition representing a task is stereotyped «SaStep».

P18—For each task the execution time is derived by making the additions of the computational budgets of each step allocated to the task. The value of this execution time is saved in the execTime property of the «SaStep»stereotype of the Activity-Partition representing the task.

P19—For each task, the priority is automatically assigned using the rate monotonic priority assignment algorithm [9]. The assigned priority is saved in the priority property of the «SaStep»stereotype of the ActivityPartition representing the task.

P20—Each ActivityPartition representing a task is mapped to an execution host. A reference to this execution host is saved in the host property of «SaStep»stereotype of the ActivityPartition representing the task.

P21—For CallBehaviorActions that are shared between tasks and whose code must be protected the sharedResources property of these actions, stereotyped «Sa-Step»are valuated with a reference to synchronization resource that is add-ed in the platform resources (see next section). The UML Behavior associated to the shared CallBehaviorActions are non-reentrant (isReentrant property is set to false).

P22—Schedulability analysis results for the tasks are serialized in «SaStep»stereotype properties of ActivityPartitions representing the tasks: respT for the response time and blockT for the blocking time. End-to-end response time is serialized in the endToEndT property of the «SaEndToEndFlow»stereotype.

Platform resources

The platform resources are modeled in a classifier stereotyped by «GaResources-Platform». The platform resources model is an abstraction of the underlying software and hardware platform that is needed to define the deployment of the concurrency model on the available resources.

The following resources are considered in this abstraction: (1) tasks, here called schedulable resources, along with their scheduling parameters, (2) synchronization resources used to solve contention in case of multi-threading (if necessary) and the protocol used for accessing them, (3) the processing and communication resources topology, (4) the scheduler algorithm used by the processing resources.

Each task, represented as an ActivityPartition in the concurrency model, is represented in the platform as a UML property stereotyped «SchedulableResource». The element ActivityPartition for the task is bound to corresponding UML property through the Activity Partition's meta-attribute "represents" which will reference the corresponding UML property stereotyped «SchedulableResource». Scheduling parameters for the task have to be specified. In the case a scheduling algorithm based on fixed priorities is assumed, priorities have to be specified as scheduling parameters. Scheduling parameters are specified through the schedParams property of «SchedulableResource». The value of a priority is in form of a VSL expression: schedParams=[fp(priority-value)]

Each synchronization resource is stereotyped as «SaSharedResource». The synchronizing protocol has to be specified in the protectKind property of the stereotype «SaSharedResource». The PriorityCeilingProtocol is used to protect the access to these critical section regions.

Table 6.2 Conformance on sets

Conformance property	Formal model	Optimum model
Event conformance	$E = (e_1, e_2, ..., e_n)$	Set of AcceptEventActions in the Activity Diagram stereotyped «GaWorkloadEvent»(property P10)
Task conformance	$T = (\tau_1, \tau_2, ..., \tau_m)$	In the Optimum model, each task corresponds to one ActivityPartition stereotyped «SaStep»and representing a UML property stereotyped «SchedulableResource»(P13)
Resource conformance	$P = (r_1, r_2, ..., r_l)$	Set of UML properties stereotyped «SaExecHost»/«SaCommHost», properties of the UML Class stereotyped «GaResourcePlatform»

The processing resources are modeled with UML Properties stereotyped «SaExecHost»and typed with the «HardwareProcessor»Class used in the workload model for the definition of the computation time budgets of functions. These processing resources are interconnected with communication resources that are modeled with UML Connectors stereotyped «SaCommHost».

The offered scheduling algorithm is specified for the processing and communication hosts. The scheduling algorithm is specified through the property schedPolicy. The value of schedPolicy is equal to FixedPriority.

6.4.3 Conformance to the Formal Schedulability Model

In this section we formally show in Tables 6.2, 6.3 and 6.4 the conformance of the Optimum model to the formal model of Sect. 6.3.

As for assumptions on the formal model, **Assumption 1 (Fixed-priority scheduler)** is specified by property schedPolicy in the stereotype «saExecHost»applied to processing/communication resources. The schedPolicy is equal to FixedPriority. **Assumption 2 (Priority-ceiling)** is expressed by the protectKind property of the stereotype «SaSharedResource»applied to shared re-sources. The protectKind is equal to PriorityCeilingProtocol. **Assumption 3 (CAN-like channels)** means considering a communication channel where policy on messages to be sent/received is based on fixed-priorities. The assumption implies thus to have each «saCommHost»with schedPolicy equal to FixedPriority. As for the **Partial-order property** on the set of task T, this property is satisfied in the Optimum model by property **P15** Tables 6.2, 6.3, 6.4 and 6.5.

Table 6.3 Conformance on event occurrences characterization

Formal model	Optimum model
ti	From property P10 each AcceptEventAction is stereotyped «GaWorkloadEvent». The period is specified in the arrivalPattern property
$R_i = (T_i, \rightarrow_i), \emptyset \subset T_i \subseteq T, T = (\tau_1, \tau_2, ..., \tau_m)$	T_i is obtained as follows: For each CallBehaviorAction *cbh* belonging to the ActivityPartition stereotyped «SaEndToEndFlow»containing AcceptEventAction (ei), take the ActivityPartition stereotyped «saStep»representing a schedulable resource (τ_h) that contains *cbh*. By property P1 no fork node there exists in the Activity diagram and by property P6 each AcceptEventAction is connect to exactly one CallBehaviorAction, this ensures that for each event there exists only one path towards the final flow node. The set of CallBehaviorActions in the «SaEndToEndFlow»is then totally ordered. The order on tasks is the order established by control flows sequencing the actions and by property P15, no cycles can occur, reducing to a total order on tasks. By property P5, exactly one ActivityPartition stereotyped «SaEndToEndFlow»there exists. By property P14 at least one ActivityPartition stereotyped «saStep»representing a schedulable resource (τ_h) there exists, thus the set of tasks $T_I \neq \emptyset$
D_i	By properties P1, P5 and P6, each AcceptEventAction belongs to only one ActivityPartition stereotyped «SaEndToEndFlow»(as already shown for R_i conformance). Deadline Di corresponds to the property end2EndD of the stereotype «SaEndToEndFlow »(P11)
WR_i	By properties P1, P5 and P6, each AcceptEventAction belongs to only one ActivityPartition stereotyped «SaEndToEndFlow». WR_i corresponds to the endToEndT (P22)

6.4.4 Software Architecture Exploration Phase

Architecture exploration aims at finding an architecture model satisfying timing requirements of the system and possibly other designer criteria. Many techniques can be employed in order to explore the design space, but here we are only interested in supporting some comparative process in which different architecture models can be built and co-habit for comparison. The key MARTE concept to support architectural exploration is the «SaAnalysisContext»concept used in P16. Different UML packages stereotyped «SaAnalysisContext»can be organized in order to store different candidates. Let us remind that the stereotype «SaAnalysisContext»allows making reference to a platform and workload, which can be different for each context. Interestingly this stereotype has an attribute optCriterion used to annotate the context with designer criteria (optimization objectives) and their weights used for the context. When the designer is satisfied by one candidate, the exploration phase terminates and the candidate architecture is given as output to the process.

Table 6.4 Conformance on task execution characterization

Formal model	Optimum model
P_h	«SaStep»stereotype, attached to the ActivityPartitions representing schedulable resources, contains the property *schedParam:SchedParameters[0..*]=[fp(priority-value)]*
$r \in P$	The «SaStep»stereotype, attached to the ActivityPartitions representing schedulable resources, contains the property host by property P20
C_i	The «SaStep»stereotype, attached to the ActivityPartitions representing schedulable resources, contains the property execTime by property P18
CS_h	The list of critical sections the task pass through is specified in the property sharedResources of the stereotype «SaStep »applied to the CallBehaviorActions belonging to the ActivityPartition representing the task by property P21
c_{hi}	The duration of the critical section csi is equal to the execution time of the CallBehaviorAction that includes the csi value in sharedResources of the applied «SaStep».
B_h	«SaStep»stereotype, attached to the ActivityPartitions representing schedulable resources, contains the property *blockT* (P22)
WTR_h	«SaStep»stereotype, attached to the ActivityPartitions representing schedulable resources, contains the property *respT* (P22)

6.5 Application on an Automotive Case Study

In this section we will illustrate the application of the methodology on an automotive case study. This subsystem is a sensor-controller-actuator system com-posed of the following functions: a data processing function for data coming from the sensor, the anti-locking brake function calculating the command to send to the actuator, and a diagnosis function that disables the anti-locking function in case a fault in the subsystem is detected.

Figure 6.3 below presents the functional model describing the system end-to-end scenarios. Each function has an associated behavior modeled as a UML Activity which is referenced by a CallBehaviorAction. Two events (acquisitionForAbs and acquisitionForDiagnosis) are triggering the sequences of functions behavior execution. Between these events and the final flow node, there are respectively two end-to-end timing requirements of 60 ms and 100 ms.

6.5.1 Workload Model

The set of system's end-to-end computations, called workload behavior, is represented with a UML activity diagram stereotyped with MARTE «GaWorkload-Behavior»stereotype. This workload behavior actually represents the behaviors activation graph. Each behavior is specified by a UML CallBehaviorAction annotated with MARTE «SaStep»stereotype. Each step has a computational budget (execTime

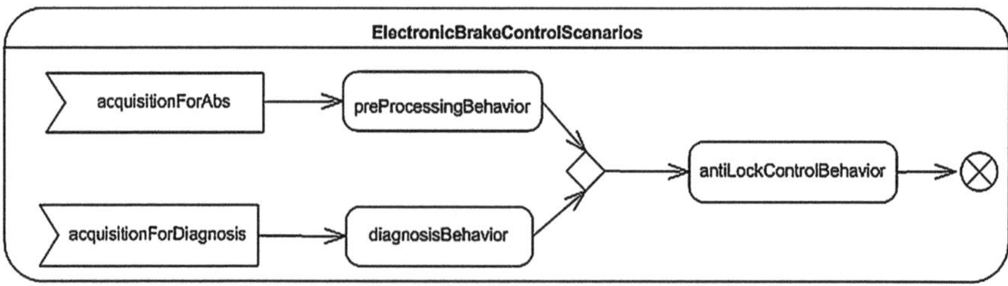

Fig. 6.3 System-level end-to-end scenarios

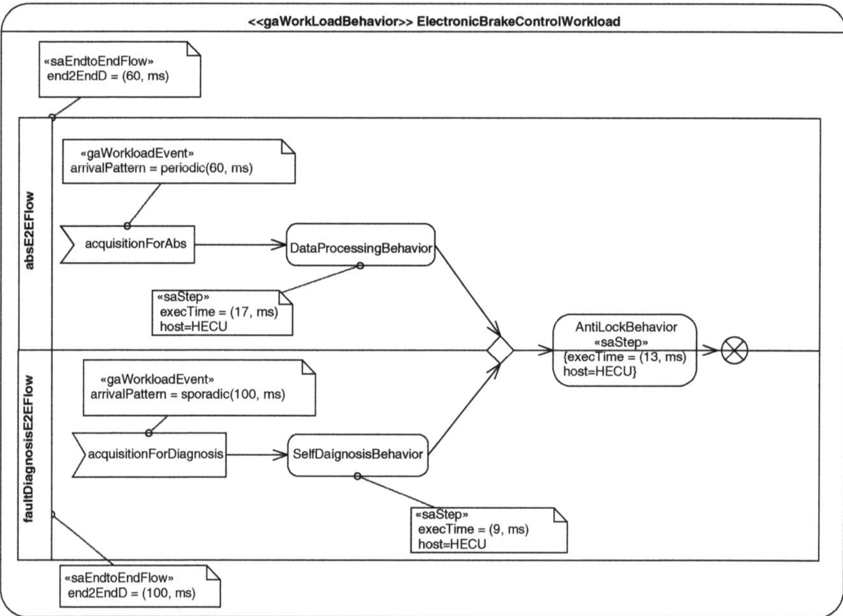

Fig. 6.4 Workload behavior

property) specified for a given type of execution host (host property). A step can be linked to a successor step with a control flow.

The workload behavior also specifies external events that trigger the steps. Each external event is modeled with a UML AcceptEventAction stereotyped «GaWorkloadEvent». The arrival pattern of the event is specified in the arrivalPattern property. For instance on Fig. 6.4, aquisitionForAbs event is periodic with a period of 60 ms.

Once that control flow graph is defined, the steps and their respective activation events are grouped in so called end-to-end flows. An end-to-end flow is modeled with a UML activity partition stereotyped «SaEndToEndFlow». An end-to-end deadline can be specified for each end-to-end flow in the end2EndD property of the «SaEndToEndflow»stereotype. The absE2EFlow has an end-to-end deadline equal to 60 ms. Let us to note that AntilockBehavior is a shared step between the two specified end-to-end flows.

As explained previously, at this stage of the methodology, we needed to define a type of processing resource in order to specify an estimation of computation time budgets for steps. This estimation can be used to perform feasibility tests with respect to expressed end-to-end deadlines and external events activation rates. The processing resource type is modeled with a UML class stereotyped «HwProcessor». In this example, HECU is the type of execution host for the steps. This type will be used in the architecture model to type the platform processing resources.

6.5.2 Generation of the Architecture Model

The following subsections give details of the generation of the Architecture Model starting from the Workload Model.

6.5.2.1 Mono-Processor Platform Definition

The proposed generation produces a mono-processor architecture. The platform (SaResources Class) includes then the property hecu of the type HECU. Inside the property host of the stereotype «saStep»applied on CallBehaviorActions, the hecu value is set. For the execution host hecu the FixedPriority scheduling algorithm is set.

6.5.2.2 Independent Tasks Generation: Protecting Shared Functions

In order to apply a scenario-based task model generation (to get one single thread of execution for each event), the situation of a function belonging to two different end-to-end flows turning at two different rates and with different dead-lines must be handled. If we map the function on two different threads (one per scenario), schedulability analysis will compute an additional blocking time necessary to protect the function code duplicated in the two threads. In fact, to avoid inconsistencies the function virtually shared by two threads has to be accessed in mutual exclusion. Blocking time for the access at the critical section (shared function code) is thus computed.

The transformed graph, therefore, will explicitly describe synchronization re-sources used to treat contention in this case. In our example the step Antilock-Behavior, originally shared between the two specified end-to-end flows, is here mapped into two different threads, namely task1 and task2. The synchronization among the two steps preceding AntilockBehavior, i.e., DataProcessingBehavior and SelfDiagnosisBehavior is modeled through the presence of a synchroniza-tion resource here named AntiLock, appearing in the property sharedResources, which actually represents a critical section for the execution of AntilockBehavior. In our example a UML property AntiLock of type SharedResource is stereotyped as

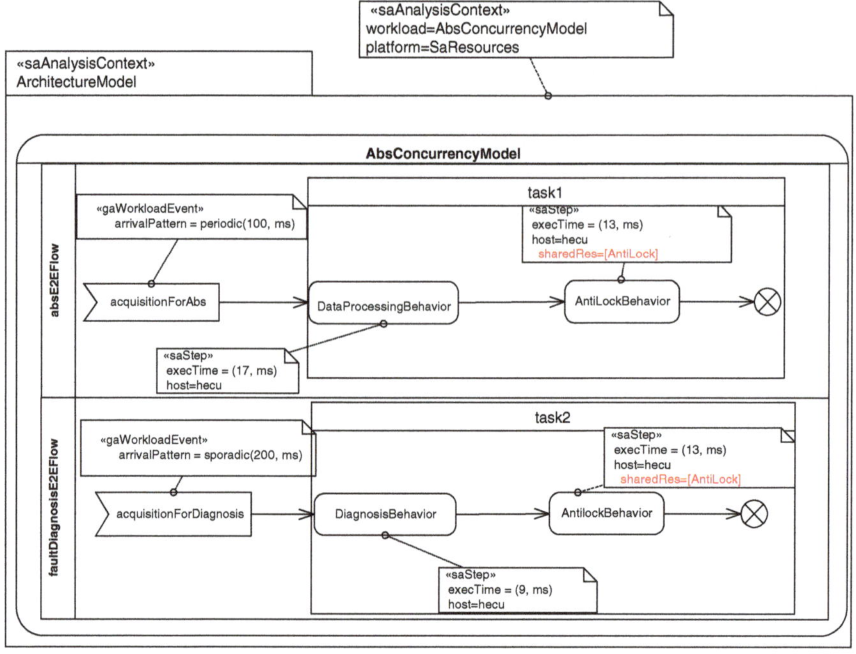

Fig. 6.5 Task mapping

«saSharedResource», and then it is included in SAResources and the synchronization protocol specified (Fig. 6.5).

6.5.2.3 Rate-Monotonic Priority Assignment

In the example the two threads task1 and task2 are included in the platform resources for schedulability analysis SaResources along with priorities. Priorities are assigned as inversely proportional to deadlines following a rate-monotonic priority assignment [11] that is optimal in case of independent tasks and fixed-priority.

6.5.3 Schedulability Analysis Results

The Optimum model as defined in the previous subsection contains all the needed information to perform schedulability analysis. Let us note that in this case responses are constituted by one single task, tasks are independent (no order) but contain a critical section with duration equal to the execution time of the Control action (13 ms). A schedulability analysis test can be carried out on this model. Such test calculates

Table 6.5 Schedulability analysis results

Task	Bi	WRi
τ_1	13	43
τ_2	0	52

worst case response times for each event. Note that the worst-case response time include the blocking time B_i calculated by the test.

Note that in this case as illustrated on Table 6.5, the highest priority task experiences a blocking time. In fact in the worst case, the lower priority task (which has a computational time lower than the highest priority task) acquires the lock before the highest priority task. Only after the lock is released (after 13 ms), task τ_1 continues its execution. For task τ_2, the worst case response time takes into account, the time for the highest priority task to execute (30) and its execution time (22).

These results can be satisfactory for the designer, but he can also evaluate other alternative architectures. In that case, multiple analysis contexts can be defined. Each analysis context will contain the evaluated architecture completed by schedulability results. Among these analysis contexts, the designer will choose the one that better satisfies its own criteria.

6.6 Related Works

In order to support the development of real-time applications a wide number of methodologies have been proposed for early analysis of non-functional requirements. While a vast number of model-based approaches have been proposed for performance prediction [12], methodologies for schedulability prediction are more recent and are gaining a growing interest with the increasing complexity of embedded real-time systems [1, 8, 13]. COMET [2] proposes a methodology for the development of concurrent and distributed real time systems but does not directly deal with the issue of defining a methodology for schedulability validation. In Saksena et al. [14] a methodology for schedulability validation of object-oriented models is proposed. The methodology starts from a design model where specification of active and passive objects, message semantics and object interaction is available. Two threading strategies are proposed: a single threading solution and a multi-threading solution. Unfortunately, while the single threading solution is analyzable and applicable, the multi-threading solution is difficult to analyze or inapplicable [5]. The problem with schedulability analysis at the design level is that a non schedulability aware design could have a concurrent model too difficult to analyze or for which no automated support exists. This problem is also shared by other methodologies such as [8, 15]. In [8], for instance, the UML-based methodology envisages and supports the use of task mapping algorithms and edf schedulability tests developed in [5] but there is no automated support for the test which is ad-hoc. In [16], the author has explored the usage of MARTE to perform schedulability analysis with the MAST tool. In this work MARTE has been used to build a MAST model library to build MAST-specific analysis models. Unlike this approach, our approach proposes methodological rules to build schedulability analysis models which are independent from any schedulability analysis tool.

6.7 Conclusions and Future Work

A lot of work has been achieved in formal approaches for timing analysis during the past decades. Despite the importance of applying timing analysis for real-time systems development, the formal nature of these approaches is an obstacle for their adoption by the software engineering community. This chapter presented a UML/MARTE front-end for these formal timing analysis approaches, focusing on the schedulability aspect and integrated in the software life cycle since the very beginning. The methodology is fully implemented in the UML modeling tool Papyrus [17]. Bridges to the MAST [18] and Rt-Druid [19] tools are integrated for schedulability tests. Let us remark that the methodology has been successfully applied in the automotive domain in the context of two collaborative projects, the European INTERESTED project [20] and the French national EDONA [21] project.

References

1. Chen, R., Sgroi, M., Lavagno, L., Martin, G., Sangiovanni-Vincentelli, A., Rabaey, J.: Uml and platform-based design pp. 107–126 (2003)
2. Gomaa, H.: Designing Concurrent, Distributed, and Real-Time Applications with Uml, 1st edn. Addison-Wesley Longman Publishing Co., Inc., Boston (2000)
3. Phan, T.H., Gerard, S., Terrier, F.: Languages for system specification. In: Real-time system modeling with ACCORD/UML methodology: illustration through an automotive case study, pp. 51–70. Kluwer Academic Publishers, Norwell, MA (2004) http://dl.acm.org/citation.cfm?id=1016425.1016431
4. Douglass, B.P.: Real Time UML: Advances in the UML for Real-Time Systems, 3rd edn. Addison Wesley Longman Publishing Co., Inc., Redwood City (2004)
5. Bartolini, C., Lipari, G., DiNatale, M.: From functional blocks to the synthesis of the architectural model in embedded real-time applications. In: Proceedings of the 11th IEEE Real Time on Embedded Technology and Applications Symposium, RTAS '05, pp. 458–467. IEEE Computer Society, Washington, DC, USA (2005). http://dx.doi.org/10.1109/RTAS.2005.24
6. Object Management Group: UML profile for modeling and analysis of real-time and embedded systems (MARTE), version 1.1, formal/2011-06-02 (June 2011). http://www.omg.org/spec/MARTE/1.1/
7. Object Management Group: UML profile for schedulability, performance, and time (spt), formal/2005-01-02 (2005). http://www.omg.org/spec/SPTP/
8. Bartolini, C., Bertolino, A., De Angelis, G., Lipari, G.: A uml profile and a methodology for real-time systems design. In: Proceedings of the 32nd EUROMICRO Conference on Software Engineering and Advanced Applications, EUROMICRO '06, pp. 108–117. IEEE Computer Society, Washington, DC, USA (2006). http://dx.doi.org/10.1109/EUROMICRO.2006.14

9. Palencia, J.C., González Harbour, M.: Schedulability analysis for tasks with static and dynamic offsets. In: Proceedings of the IEEE Real-Time Systems Symposium, RTSS '98, pp. 26-. IEEE Computer Society, Washington, DC, USA (1998). http://dl.acm.org/citation.cfm?id=827270.829048
10. Lehoczky, J.P., Sha, L., Ding, Y.: The rate monotonic scheduling algorithm: Exact characterization and average case behavior. In: RTSS, pp. 166–171 (1989)
11. Audsley, N.: Optimal Priority Assignment And Feasibility Of Static Priority Tasks With Arbitrary Start Times. Department of Computer Science, University of York, England, Tech. rep. (1991)
12. Balsamo, S., Di Marco, A., Inverardi, P., Simeoni, M.: Model-based performance prediction in software development: A survey. IEEE Trans. Softw. Eng. **30**(5), 295–310 (2004). http://dx.doi.org/10.1109/TSE.2004.9
13. Jin, D., Levy, D.C.: An approach to schedulability analysis of uml-based real-time systems design. In: Proceedings of the 3rd international workshop on Software and performance, WOSP '02, pp. 243–250. ACM, New York, NY, USA (2002). http://doi.acm.org/10.1145/584369.584409
14. Saksena, M., Karvelas, P.: Designing for schedulability: integrating schedulability analysis with object-oriented design. In: Proceedings of the 12th Euromicro conference on Real-time systems, Euromicro-RTS'00, pp. 101–108. IEEE Computer Society, Washington, DC, USA (2000). http://dl.acm.org/citation.cfm?id=1947412.1947431
15. Gu, Z., He, Z.: Real-time scheduling techniques for implementation synthesis from component-based software models. In: Proceedings of the 8th international conference on Component-Based Software Engineering, CBSE'05, pp. 235–250. Springer, Berlin (2005)
16. Kenneth, J.: Schedulability analysis of embedded applications modelled using marte. Ph.D. thesis, Technical University of Denmark, DTU, DK-2800 Kgs. Lyngby, Denmark (2009)
17. Eclipse Foundation: Papyrus (2010). http://www.eclipse.org/modeling/mdt/papyrus/
18. Pasaje, J.L.M., et al.: UML-MAST (2005). http://mast.unican.es/umlmast/
19. Evidence: Home page (2009). http://evidence.eu.com
20. Interested, EU Project ICT-2007-3.3 B,: Project web page (2009). http://www.interested-ip.eu/
21. Edona, P.: Environnements de Développement Ouverts aux Normes de l'Automobile (2009). http://www.edona.fr

Part III
Component-Based Design and Real-Time Components

Chapter 7
Early Time-Budgeting for Component-Based Embedded Control Systems

Manoj G. Dixit, S. Ramesh and Pallab Dasgupta

Abstract One of the challenging steps in the development of component based embedded control systems involves decomposition of feature or system level timing requirements into component level timing requirements. Often it is observed that the timing is introduced at a later stage in the development cycle and ad hoc estimates are made which lead to costly and multiple design iterations. This chapter proposes a methodology that addresses this problem using a simple but powerful idea of using parametric specification. A key step in the methodology is *component time-budgeting*, which involves identifying a set of parametric timing requirements for the components realizing a feature functionality. This is followed by a verification step which computes a set of constraints on the parameters such that any valuation of the parameters satisfying the constraints achieves the feature requirements. This avoids the ad hoc time estimates and the consequent design iteration. The methodology is formalized using Parametric Temporal Logic and illustrated on a reasonably sized automotive case study.

7.1 Introduction

Modern automobiles and aircrafts are complex embedded control systems having hundreds of features like collision mitigation [1] and Brake By-Wire [2]. The software engineering methodologies for such real-time embedded systems have

M. G. Dixit (✉) · S. Ramesh
India Science Lab, GM R&D, Bangalore, India
e-mail: mgdixit@gmail.com

S. Ramesh
e-mail: ramesh.s@gm.com

P. Dasgupta
I.I.T Kharagpur, Kharagpur, India
e-mail: pallab@cse.iitkgp.ernet.in

A. Sangiovanni-Vincentelli et al. (eds.), *Embedded Systems Development*, Embedded Systems 20, DOI: 10.1007/978-1-4614-3879-3_7,

recommended hierarchical decomposition techniques for implementing features using multiple components [3–8]. In our view, there are two distinct steps found in these methodologies: *functional decomposition* and *component implementation*. The functional decomposition identifies components and their requirements for achieving feature requirements.The component implementation step involves mapping component functionality to a set of tasks which are executed over one or more computing nodes in a distributed network. The component-node allocation [9], component-task mapping [10], task-period synthesis [11] and schedule definition [12] are some problems that get addressed during this step.

To manage the increasing complexity and scalability issues, there is a further need to refine the above steps and bring verification and validation flow much earlier in the development phase. This poses several challenges in specifying the given system at right level abstraction so that formal reasoning can be done. Specifically, we focus on component timing requirements and observe that due to lack of effective methods for identifying them at an early stage of functional decomposition, ad hoc estimates are made in practice. This usually leads to costly design iterations.

In order to bridge the above gap, we have proposed an early stage component time-budgeting methodology. A salient feature of this methodology is the use of *parametric timing* in the requirement specifications. As shown in Fig. 7.1, we have enhanced the present functional decomposition step to identify the component level parametric timing requirements from feature level real-time requirements. This is followed with a verification step which computes a set of constraints from such parameterized specifications. The computed constraints provide guarantee that any common solution provides the values for the parameters so that the component requirements instantiated with these parameter values imply the feature requirements. The computed constraints are then used in the design space exploration stage. At this stage, application specific additional constraints can be placed on the parameters and a suitable optimization criteria can be selected. The combined set of constraints can be optimized with respect to the selected criteria to obtain a parameter valuation. We observe that separating out constraints in this way provides significant flexibility in selecting suitable concrete component specifications for the later stages of development.

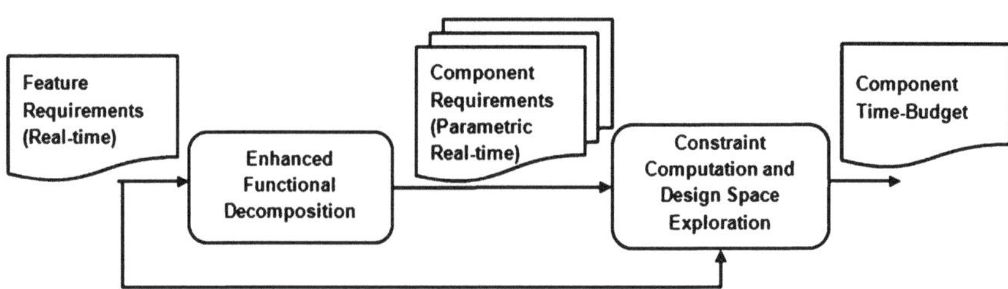

Fig. 7.1 Component time-budgeting methodology

In this chapter we have presented the above methodology by using Parametric Temporal Logic (PLTL) [13] for specifying requirements. PLTL is suitable for specifying qualitative and symbolic temporal relationships between important events of the given system. This provides a right level of abstraction for specifying early stage requirements. We reduce the component time-budgeting computation to checking validity of parametric PLTL formulae and used some of the interesting properties of PLTL for computing the constraints. A preliminary version of this methodology has been presented in Refs. [14, 15]. A more comprehensive description including case studies is available in Ref. [16].

7.1.1 Related Work

In the traditional development, the configuration of timing values is usually restricted to the attributes of tasks/messages [10, 12, 17, 18]. Our work focuses on early phase of requirements engineering and on configuring timing bounds in component timing requirements. The automotive standard AUTOSAR [19] has proposed the use of TADL specification language and TIMMO methodology [20] for an early modeling and analysis of timing requirements. The primary difference as we see between the two approaches is the use of parameters to represent the component level timing. The constraints computed in our approach provide more flexibility in selection of suitable values for timing constants. TADL on the other hand, is a domain specific language and suitable for specifying architectural level timing properties. Temporal logics have always been a natural choice as a specification language for real-time embedded systems [13, 21–25]. The time-budgeting approach outlined in this chapter uses temporal logic PLTL [13] for specifying feature level real-time and component level parametric real-time requirements. In this work, we have provided more scalable algorithms for computing a constraint based representation of the validity set for a given parametric formula than the solutions given in Ref. [13].

The rest of the chapter is organized as follows: Sect. 7.2 formalizes the component time-budgeting methodology, Sect. 7.3 describes the constraint computation methods for parameterized specifications, Sect. 7.4 describes the results on some of the case studies and Sect. 7.5 concludes by providing future directions for research.

7.2 Time-Budgeting Methodology

At the broad level, the time-budgeting methodology takes the following (extended) functional decomposition as its inputs:

- A set of features and their real-time requirements,
- A component hierarchy implementing these features along with the parametric-time requirements identified for each component while decomposing the feature requirements, and

- Add-on constraints and optimization criteria over component parameters to constrain their values.

The methodology outputs a parameter valuation called *component time-budget* such that component requirements when instantiated with this valuation together satisfy feature requirements. In the next subsection, we have formalized all inputs and defined a component time-budget based on it. In the subsequent subsection, the component time-budget computation procedure is given.

7.2.1 Formalization of Component Time-Budgeting

For specifying requirements, we have used Parametric Temporal Logic (PLTL) proposed by Alur et al. [13]. PLTL extends the well known Linear Temporal Logic (LTL) [25] operators with quantitative and parametric bounds over the set $\mathbb{N}$ of non-negative integers on event occurrences. Given a set P of atomic propositions and two disjoint sets X, Y of parameter variables, let $p \in P,\ x \in X \cup \mathbb{N},\ y \in Y \cup \mathbb{N}$. PLTL formulae are defined using the following grammar[1]:

$$\phi := p \mid \neg\phi \mid \phi \vee \phi \mid \phi \wedge \phi \mid \mathcal{O}\phi \mid \Box\phi \mid \Diamond\phi \mid \phi\ \mathcal{U}\ \phi \mid \phi\ \mathcal{R}\ \phi \mid \Box_{\leq y}\phi \mid \Diamond_{\leq x}\phi$$

$\vee, \wedge, \neg$ are standard propositional connectives; $\Box$ ('Globally'), $\Diamond$ ('Eventually'), $\mathcal{U}$ ('Until'), $\mathcal{R}$ ('Release'), $\mathcal{O}$ ('Next') are temporal operators.For example, the PLTL formula $\Box_{\leq 10} p$ requires the event p to occur continuously for 10 units of time, whereas, the PLTL formula $\Diamond_{\leq x} q$ requires that the event q should occur within x units of time. Let $\alpha : X \cup Y \rightarrow \mathbb{N}$ be a parameter valuation. A PLTL formula Φ is satisfiable with respect to α *iff* the parameter-free formula $\Phi(\alpha)$ obtained by substituting $z \in X \cup Y$ with $\alpha(z)$ is satisfiable. For checking the satisfiability of $\Phi(\alpha)$, a simple extension of the standard semantic rules for LTL is used. PLTL$_\Box$ is a fragment of PLTL in which parameter variables are associated only with $\Box_{\leq y}$ and PLTL$_\Diamond$ is a fragment of PLTL in which parameter variables are associated only with $\Diamond_{\leq x}$. A requirement pattern that is of particular interest to this work is the *bounded-response* pattern. A bounded-response formula has the form $\Box(\phi \Rightarrow \Diamond_{\leq x}\psi)$, where, ϕ and ψ are Boolean formulae and $x \in X \cup \mathbb{N}$.

The component hierarchy for the given feature set is specified by a Directed Acyclic Graph (DAG), in which the root nodes (i.e., those without incoming edges) correspond to features, whereas, the remaining ones represent components. We specify the feature requirement decomposition across the component hierarchy by a collection of *Requirement Decomposition Pairs* (RDPs). An RDP is a pair $(f, \{g_1, \ldots, g_n\})$ such that $g_1, \ldots, g_n$ are the requirements identified while decomposing the requirement f. Note that these requirements correspond to feature/ components in the successive levels of the hierarchy.

---

[1] The logic given in [13] has formulae only in *negation normal form* ('$\neg$' precedes only members of P), whereas, we use '$\neg$' with arbitrary formulae. However, two logics are equivalent.

Add-on constraints are standard linear constraints over component parameters. An optimization criterion is an objective function like maximizing or minimizing parameter-sum for selecting a suitable valuation of parameters. This criterion is specified at each successive levels of decomposition in the hierarchy.

Example 1 Fig. 7.2a shows the component decomposition DAG for two automotive features: Adaptive Cruise Control (ACC) and Collision Mitigation (CM). ACC provides control functionality to maintain vehicle at driver set-speed or set-gap from a lead vehicle. CM provides functionality to mitigate intensity of collision through braking. All internal nodes are components implementing these two features. For example, *Sensor Subsystem* provides functionality for lead vehicle detection. This component is further divided into three subcomponents: *Radar*, *Dynamics* and *Data Fusion*. The table in Fig. 7.2b shows an illustrative example of decomposition of ACC feature requirement ϕ_1 into component level requirements ψ_1 and ψ_2. This decomposition is described by an RDP $(\phi, \{\psi_1, \psi_2\})$. Note that the time-bound is specified by a constant (500) in the feature requirement, whereas, it is parametric (x_1, x_2) in component requirements. ■

For the given functional decomposition, a component time-budget is defined as follows:

(a)

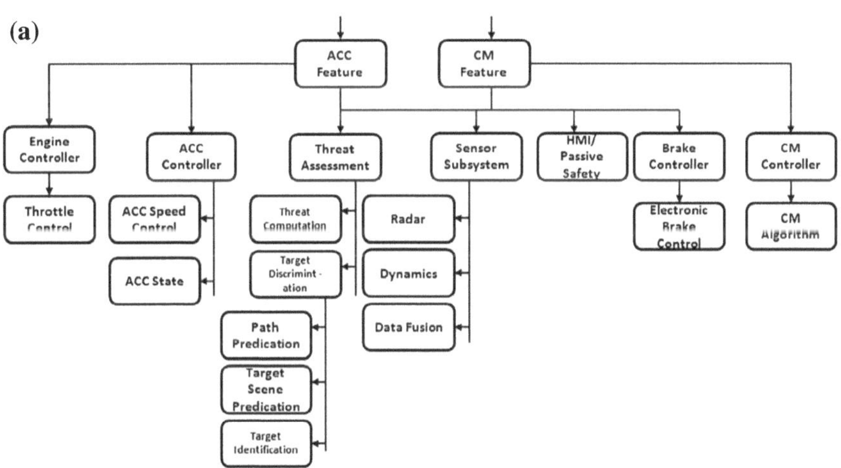

(b)

Feature/Component Name	PLTL Formula
ACC Feature	$\phi_1 : \Box(lead\_slow \Rightarrow \Diamond_{\leq 500}\, apply\_brake)$
Sensor Subsystem	$\psi_1 : \Box(lead\_slow \Rightarrow \Diamond_{\leq x_1}\, lead\_kinematics\_info)$
ACC Controller	$\psi_2 : \Box(lead\_kinematics\_info \Rightarrow \Diamond_{\leq x_2}\, apply\_brake)$

Fig. 7.2 Functional decomposition of ACC feature

Definition 1 *(Component Time-budget)* A component time-budget is a parameter valuation α such that for any RDP $(\Phi, \{\Psi_1, \ldots, \Psi_n\})$ in the given functional decomposition, the parameter-free formula $(\bigwedge_{i=1}^{n} \Psi_i \Rightarrow \Phi)(\alpha)$ is valid. ■

Observe that there may be many choices for a component time-budget. For example, if we restrict the functional decomposition to the requirements given in the table from Fig. 7.2b, any solution satisfying $x_1 + x_2 \leq 500$ is a component time-budget.

7.2.2 Component Time-Budget Computation

For a given functional decomposition, we now describe a method for the component time-budget computation. At the broad level, this method involves traversing the given component hierarchy and incrementally deriving parameter values resulting in a component time-budget. For example, the given feature level timing constants are first used in finding a suitable valuation of parameters defined for all top level components. These values are subsequently used in finding a suitable valuation of parameters defined for their sub-components and so on.

As discussed in the previous subsection, there may be various choices for a suitable parameter valuation which results in a component time-budget. Hence, to provide necessary flexibility, our method involves computing a constraint viz. *requirement decomposition constraint* (RDP constraint) corresponding to each RDP in the given functional decomposition. Formally,

Definition 2 *(Requirement Decomposition Constraint)* A constraint $C(X, Y)$ is a requirement decomposition constraint for an RDP $(\Phi, \{\Psi_1, \ldots, \Psi_n\})$ provided for any valuation α satisfying $C(X, Y)$, the parameter-free formula $(\bigwedge_{i=1}^{n} \Psi_i \Rightarrow \Phi)(\alpha)$ is valid. ■

For simplicity of description below, we assume that component hierarchy involved in the functional decomposition is defined for a single feature. The component level time-budgeting method is a semi-decision procedure involving traversal of the DAG for the given component hierarchy in the depth-first fashion starting from the feature node. At each non-leaf component node in the traversal, as shown in Fig. 7.3, the Time Budgeting Step uses the following two sub-steps to compute a suitable valuation of parameters defined for its child node components:

- The *Requirements Decomposition Constraint Extractor* sub-step computes RDP constraints for all RDPs identified for the next-level functional decomposition of the given component node. It outputs *Requirement Decomposition Constraint* as a conjunction of all computed RDP constraints.
- The *Design Space Explorer* sub-step combines the requirement decomposition constraint with add-on constraints defined by the user and outputs a parameter valuation satisfying the optimization criteria for the given node. Observe that this defines a valuation of all parameters used in the requirements of child nodes of the

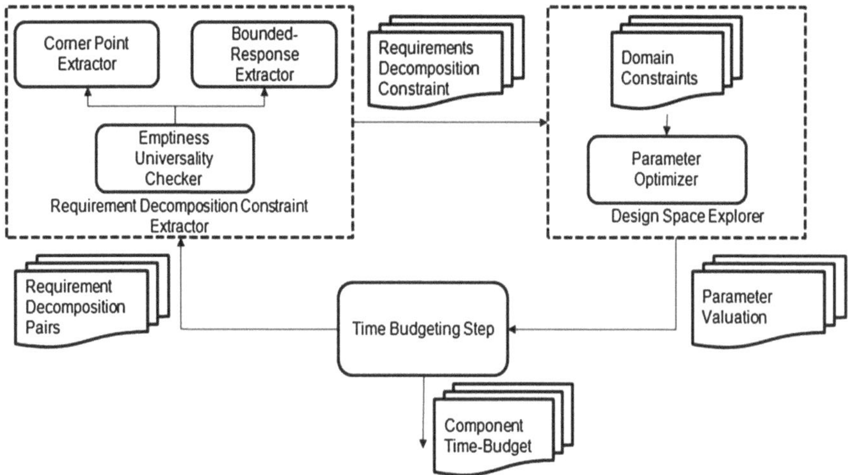

Fig. 7.3 Computing parameter valuation at a node in DAG

node under processing. The child node timing requirements are instantiated with the corresponding parameter values before advancing the traversal.

The requirement decomposition constraint computed above, may be unsatisfiable in some cases. In such cases, the traversal is back-tracked and the design space exploration sub-step is repeated at the parent node to select another valuation so that traversal can be advanced. It is easy to check that when DAG traversal is complete, the resulting parameter valuation is a component time-budget.

We now discuss the important sub-step of computing an RDP constraint in the time-budgeting algorithm. Given an RDP, we compute the RDP constraint in the form of a *linear predicate*. Formally,

Definition 3 *(Linear Predicate)* A linear predicate is either a linear inequality or a Boolean combination of linear inequalities. ■

The required RDP constraint computation is achieved in the following two steps:

1. For an RDP, *Emptiness and Universality Checker* finds out whether the corresponding RDP constraint is trivial or not.
2. If the RDP constraint is found to be non-trivial in the previous step, then:
   - If all formulae involved in the given RDP are in bounded response form, then *Bounded Response Extractor* computes the required RDP constraint.
   - F or an RDP having more general formulae, *Corner Point Extractor* computes the required RDP constraint.

In the next section we describe the details of the constraint computation methods used by the above constraint extractors.

7.3 RDP Constraint Computation Methods

Given an RDP $(\Psi, \{\Psi_1, \ldots, \Psi_n\})$, we present methods to compute a linear predicate as its RDP constraint. Recall that any solution to the computed linear predicate makes the RDP implication formula valid.

7.3.1 Emptiness and Universality Check Method

For the given RDP, we have defined a method to check whether the corresponding RDP constraint is trivial i.e., whether the set of valuations that make the RDP implication formula valid is empty or universal [15]. If the set of valuations that make the RDP implication valid is found to be empty, then the RDP constraint is `false`, which represents an unsatisfiable constraint. If the set of valuations is found to be universal, then the RDP constraint is $\bigwedge_{x \in X \cup Y} x \geq 0$.

Our method involves reducing the RDP implication formula $\Phi \triangleq (\bigwedge_{i=1}^{n} \Psi_i \Rightarrow \Psi)$ to a parameter-free formula and then carry out the validity check over the reduced formula. To obtain a parameter-free formula, a simple *parameter abstraction* operation has been defined for PLTL formulae. This operation reduces a PLTL formula Δ to the parameter-free formula $\widetilde{\Delta}$ by simply dropping the parameter subscripts from the temporal operators. For example, let $\Delta = \Box_{\leq y}\, p_1 \wedge \Box_{\leq 10}\, p_2$ where $p_1, p_2 \in P$. Then the parameter-abstraction of Δ is the formula $\widetilde{\Delta} = \Box\, p_1 \wedge \Box_{\leq 10}\, p_2$. Note that we do not drop the subscript '$\leq$ 10' while defining $\widetilde{\Phi}$.

Let V_Φ represent the set of all valuations that make the RDP implication formula Φ valid. 0_X and 0_Y are (partial) parameter valuations that assign the value 0 to all members of the sets X and Y respectively. We are now ready to describe the solution for emptiness and universality checks. For PLTL fragments and emptiness/universality problems, the entries in Table 7.1 contain parameter-free formulae derived from the given RDP implication formula Φ. The row entries are classified based on whether Φ is equivalent to a formula in PLTL$_\Box$, PLTL$_\Diamond$ fragments (Rows 2 and 3 respectively) or not (Row 4). For example, the entry, $\Phi(0_X)$, corresponding to 'PLTL$_\Diamond$ fragment'[2] and 'V_Φ universality' is obtained by substituting all members of the parameter set X in Φ with the constant 0. For each fragment, the set V_Φ is

Table 7.1 Parameter-free formulae for emptiness and universality checks

	V_Φ is Non-Empty	V_Φ is Universal
PLTL$_\Box$	$\Phi(0_Y)$	$\widetilde{\Phi}$
PLTL$_\Diamond$	$\widetilde{\Phi}$	$\Phi(0_X)$
PLTL	$\widetilde{\Phi(0_Y)}$	$\widetilde{\Phi(0_X)}$

[2] Recall for PLTL$_\Diamond$ formulae, parameters from the set X are only used.

non-empty whenever the corresponding formula in Column 2 is valid. For example, in case of PLTL$_\Box$ fragment, V_Φ is non-empty when $\Phi(0_Y)$ is valid. In the similar way, the set V_Φ is universal whenever the corresponding formula in Column 3 is valid. For example, in case of PLTL$_\Box$ fragment, V_Φ is universal when $\widetilde{\Phi}$ is valid.

7.3.2 Bounded Response Constraint Extraction Method

We consider the constraint computation case when the given RDP $(\Psi, \{\Psi_1, \ldots, \Psi_n\})$ contains only bounded-response formulae. A bounded-response formula $\Box(\phi \Rightarrow \Diamond_{\leq x}\psi)$ has two main Boolean subformulae viz. antecedent, ϕ, and consequent, ψ. For a bounded-response formula Φ, we represent antecedent and consequent subformulae by Φ^a and Φ^c respectively.

Our constraint computation method starts with the set of antecedent and consequent subformulae obtained from the given RDP and constructs a rooted tree. The tree consists of three types of nodes: *property*-nodes labeled with one of the formulae in the given RDP and two special types of nodes viz. *and*-node and *or*-node. Once the tree is constructed, the linear predicate computation involves assigning a linear predicate to each node in the tree.

In the tree construction process, we have used a notion of an *irreducible cover* for a collection of Boolean formulae. Specifically, we are given a Boolean formula γ and a set $\{\gamma_1, \ldots, \gamma_n\}$ of Boolean formulae. An irreducible cover of γ is a minimal subset H of $\{\gamma_1, \ldots, \gamma_n\}$ such that the Boolean implication formula $\gamma \Rightarrow \bigvee_{\gamma' \in H} \gamma'$ is valid. For example, for the formula $p \vee q$, $p, q \in P$ and the set $\{p, q\}$, there is a single irreducible cover viz. $\{p, q\}$, whereas, for the formula $p \wedge q$ with the same set, there are two irreducible covers: $\{p\}$ and $\{q\}$.

Tree Construction: We are now ready to describe the details of tree construction for the RDP $(\Psi, \{\Psi_1, \ldots, \Psi_n\})$. The root of the tree is a *property*-node labeled Ψ. The tree expansion happens only at a *property*-node provided the following conditions are met: (a) the RDP formula labeling the node is different from formulae labeling any of its predecessor *property*-nodes and (b) the consequent of the RDP formula labeling the node does not logically imply consequent of Ψ. The expansion involves the following steps:

1. Let the given *property*-node be labeled with the formula Δ from the RDP. We consider a Boolean formula γ as follows: If the *property*-node is the root, then γ is $\Delta^a \wedge \neg\Psi^c$, otherwise γ is $\Delta^c \wedge \neg\Psi^c$. We compute all irreducible covers of γ defined by the set $\{\Psi_1^a, \ldots, \Psi_n^a\}$. If at least one irreducible cover exists, then one *or*-node is added as a child of the current *property*-node.
2. Let H be an irreducible cover obtained in the previous step. Then we define $H' \subseteq \{\Psi_1, \ldots, \Psi_n\}$ such that $\delta \in H'$ *iff* $\delta^a \in H$.
3. For every set H' obtained in the previous step, one *and*-node is added as a child of the *or*-node added in step 1. Additionally, for every member of H', a *property*-node labeling it is added as a child of the newly added *and*-node for H'.

Example 2 Let p, q, r and s be the atomic propositions. We consider two different definitions of the RDP $(\Phi, \{\Psi_1, \Psi_2\})$:

1. $\Phi \triangleq \Box(p \Rightarrow \Diamond_{\leq 5} r)$ and $\Psi_1 \triangleq \Box(p \Rightarrow \Diamond_{\leq x} q)$, $\Psi_2 \triangleq \Box(q \Rightarrow \Diamond_{\leq y} r)$. Fig. 7.4a shows the tree for this RDP.
2. $\Phi \triangleq \Box(p \vee q \Rightarrow \Diamond_{\leq 5} r)$ and $\Psi_1 \triangleq \Box(p \Rightarrow \Diamond_{\leq x} r)$, $\Psi_2 \triangleq \Box(q \Rightarrow \Diamond_{\leq y} r)$. Fig. 7.4b shows the tree for this RDP. ■

Constraint Computation: Once the tree is constructed for the given RDP, we associate a constraint with each node in the tree. If the tree has single node and $\Psi^a \Rightarrow \Psi^c$ is a valid formula, then the node constraint is $\bigwedge_{x \in X \cup Y} x \geq 0$, otherwise it is `false`. If the tree is non-trivial, we assign a node constraint as follows:

- *property*-node: If the *property*-node is a non-leaf node, the node constraint is same as the constraint associated with its *or*-child. In the other case, let Δ be the label of the node, if $\Delta^c \Rightarrow \Psi^c$ is a valid formula, then the node constraint is $\sum_{x \in X'} x \leq n$, where, X' is the set of parameters associated with the formulae labeling all *property*-nodes on the path from root to current node and n is the constant (or a parameter) used in the formula Ψ from the RDP. If the implication is not valid, then the node constraint is `false`.
- *or*-node: the node constraint equals *disjunction* of all node constraints associated with *and*-child nodes.
- *and*-node: the node constraint equals *conjunction* of all node constraints associated with *property*-child nodes.

We have the following result which is proved in [16].

Theorem 1 *For the given RDP, the linear predicate computed for the root node of the constructed tree is an RDP constraint.*

For example, the RDP constraint for the tree in Fig. 7.4a is $x + y \leq 5$, whereas, it is $(x \leq 5) \wedge (y \leq 5)$ for the tree in Fig. 7.4b.

7.3.3 Corner Point Constraint Extraction Method

We now discuss RDP constraint computation for the general case. We consider the RDP implication formula $(\bigwedge_{i=1}^{n} \Psi_i \Rightarrow \Psi)$ and consider the three cases depending

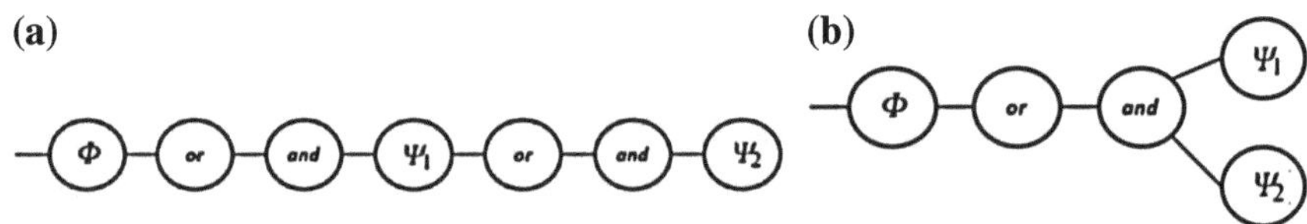

Fig. 7.4 Trees constructed for RDPs in Example 7.2 **a** Tree for (1) **b** Tree for (2)

on whether this formula: (a) belongs to PLTL$_\Box$ fragment (b) belongs to PLTL$_\Diamond$ fragment and (c) neither of the two subfragments.

For a PLTL$_\Box$ formula Φ, let V_Φ be the set of all valuations that make Φ valid. We have developed a prune and search based algorithm for computing the linear predicate based representation of the set V_Φ [14, 16]. The use of *downward closure* property satisfied by PLTL$_\Box$ formulae is central to this approach. According to this property, for a formula Φ, if $\alpha \in S$, then a valuation β, such that $\forall y \in Y, \beta(y) \leq \alpha(y)$ (i.e., α *dominates* β), also belongs to S. Based on this, our method computes all members of V_Φ which are not dominated by any other valuation. We call these as *corner points* of V_Φ. Let $\alpha_1, \ldots, \alpha_k$ be all corner points of V_Φ, then the constraint represented by them is: $\bigvee_{i=1}^{k}(\bigwedge_{y \in Y} y \leq \alpha_i(y))$. The following result ensures the correctness of the corner point based constraint computation [16].

Theorem 2 *For a PLTL$_\Box$ formula Φ, let $C(X, Y)$ be the corner point based constraint, then $C(X, Y)$ is a representation of V_Φ.*

The prune and search algorithm searches the space of parameter valuations exhaustively to find all corner points. An example of the search step for two dimensional case is shown in Fig. 7.5a. The figure shows the region below staircase consisting of valuations that make the implication formula corresponding to the RDP 1 from Example 7.2 valid. $\alpha_1, \ldots, \alpha_6$ are the corner points of this validity region. The search-step iteratively finds a corner point starting from the given base point (origin in the figure). The *iteration 1* finds the farthest member β in the validity set lying on the diagonal originating at the base point. Note that β has optimal value in at least one of x_1 or x_2. The parameter x_2 is fixed because it is optimal and the *iteration 2* is carried out in the similar way as the iteration 1 but in the reduced dimension. This results in finding the corner point α_4. The prune-step shown in Fig. 7.5b partitions $\mathbb{N}^2$ into 4 parts at α_4: $\mathscr{R}_1, \ldots, \mathscr{R}_4$. Note that regions $\mathscr{R}_1$ and $\mathscr{R}_3$ can-not contain any corner points, hence are pruned from further search. The search-step is repeated by selecting new base-points in the remaining two regions. For the k dimensional case,

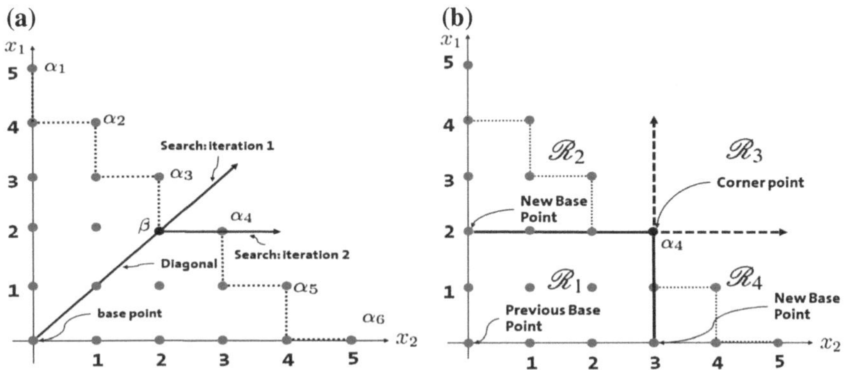

Fig. 7.5 Corner point extraction steps for RDP 1 in Example 7.2

at most k search iterations are required to reach a corner point, whereas, prune step finds at most k new base-points.

For the unbounded sets, we allow '∞' as an additional value that a parameter can have in a corner point. To locate the farthest point, we use the universality test discussed previously. We have proved a more general result that any downward closed subset of $k \geq 1$, $(\mathbb{N} \cup \infty)^k$ has only finite number of corner points [16]. This result ensures the termination of our algorithm.

Based on duality relationship [13] between $\text{PLTL}_{\Box}$ and $\text{PLTL}_{\Diamond}$, the same algorithm can be adapted for $\text{PLTL}_{\Diamond}$ fragment [16]. For the general case, it is shown that there does not exist an algorithm that given a PLTL formula Φ, computes a linear predicate over $X \cup Y$ which defines the set V_{Φ} [15] . Hence, the constraint computation method uses the heuristic of reducing the given PLTL formula to one or more formulae in $\text{PLTL}_{\Box}$ or $\text{PLTL}_{\Diamond}$ fragments by fixing one or more parameters with constants. The corner point algorithm on reduced formulae is used to compute a linear predicate which is an under-approximation of the original set V_{Φ} [16].

7.4 Case Studies

We have implemented a prototype version of our methodology in Eclipse [26], using the Java programming language, the NuSMV model checker [27] and the Yices constraint solver [28]. The experiments were carried out on a machine with Intel Centrino Duo Processor (2.2 GHz) and 2GB RAM. For demonstration, we considered the two automotive features, ACC and CM introduced earlier in Example 1.

We started with a *single-level* functional decomposition. For specifying complex temporal relationships involved in the decomposition, we needed general PLTL formulae. In the time-budgeting step, only corner-point constraint extraction method was used. In our experimentation we have found that corner points are only a small fraction of total points in the solution region. Figure 7.6a shows their distribution. The horizontal axis shows the RDP identifiers and the vertical axis shows the percentage split between non-corner points (light shade) and corner points (dark shade). The average run-time of constraint computation is observed to be increasing rapidly with search dimension leading to scalability issues.

To get around the scalability issues, we experimented with a separate *multi-level* functional decomposition shown in Fig. 7.2a. In this decomposition, all top level components had requirements in the bounded-response form. The hierarchical decomposition resulted in all leaf level components having requirements in simpler form as compared to the ones in our earlier decomposition. In the time-budgeting step all constraint extraction methods were used. The analysis time was found to be substantially improved in this case.

In all, we considered around 120+ feature and component timing requirements in our analysis. The table in Fig. 7.6b shows the performance metric for a single and multilevel component decomposition approach. The latter one is more scalable and has significantly better run-time due to the use of bounded-response pattern in

(a)

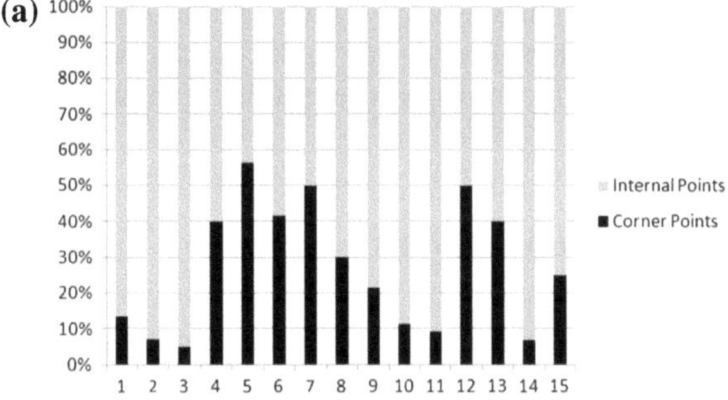

(b)

Decomposition Type	Num Components	RDP Nos	Num. SAT Checks (Average)	Constraint Computation Time (Average) (sec)
Single Level	12	15	314 (LTL)	1280
Multi Level	22	11 (Bounded-Resp), 20 (General)	83 (Boolean SAT), 57 (LTL)	2.36 (Boolean SAT), 19.25 (LTL)

(c)

Component	Parameter	Only ACC	Only CM	ACC-CM
Brake Controller Component	w_1	200	160	160
	w_2	70	-	70
	w_3	100	-	100
	w_4	100	-	100
	w_5	10	-	10
	w_6	150	100	100
Electronic Brake Control	w_{11}	35	20	20
	w_{12}	100	80	80
	w_{13}	30	20	20
	w_{14}	10	40	20
	w_{15}	100	100	100
	w_{16}	60	-	60
	w_{17}	60	-	60

Fig. 7.6 Illustrative results for ACC and CM time-budgeting **a** Corner Point-Internal Point Distribution **b** Single Level/Multi Level Decomposition Comparison **c** Illustration of Varying Time-Budgets

the requirements. Our experimental results have shown that a combination of these analysis methods supported by hierarchical decomposition will improve overall run-time of the analysis and provide scalability to handle functional decompositions found in practice.

For the design space exploration, 100+ add-on constraints were used. The add-on constraints were of two types: (a) range constraint for a parameter and (b) synchronization constraint to bind the relative difference between one or more parameters. We have additionally defined a hierarchical multi-parameter optimization scheme. In this scheme, we assign a priority to each parameter and parameter valuations are selected in the priority order. For selecting optimal valuation at each priority level, *maximizing sum of parameters* was used as an optimization criteria.

We repeated the time-budgeting step for the three feature combinations: ACC only, CM only and ACC/CM together. Figure 7.6c shows the computed parameter values for two components in the decomposition. The highlighted gray cells show the impact of feature level timing on component level timing and how component level time-budget varies with the selected feature combination. For example, when ACC is only present, the value of parameter w_1 for Brake Controller Component is 200 ms, whereas, it reduces to 160 ms, whenever CM is present.

7.5 Conclusion and Future Work

We have presented an early stage time-budgeting methodology to address the problem of decomposition of feature level timing requirements to component level timing requirements. A salient feature of this methodology is the use of parametric specifications for modeling requirements. A suitable valuation of parameters is found by computing constraints over these parameters.

The work done here can be extended in many ways: To improve scalability, specially tuned constrained computation algorithms for widely used requirement patterns are required. Another similar interesting problem to explore is to define a more efficient satisfiability checking method for parameter-free PLTL formulae. For embedded systems development, an important problem is to define an integrated methodology consisting of time-budgeting methodology and architecture exploration methods so as to refine the feature and component level timing into architectural level attributes like periods of tasks.

References

1. FMCSA: Forward Collision Warning Systems (CWS) (2005). http://www.fmcsa.dot.gov/facts-research/research-technology/report/forward-collision-warning-systems.htm
2. Plunkett, D.: Safety critical software development for a brake by-wire system. In: SAE Technical Paper Series, 01-1672 (2006)
3. Cameron, J.: An overview of JSD. IEEE Trans. Softw. Eng. **12**(2), 938–949 (1986)
4. Douglass, B.: Real Time UML. Pearson, Education, Boston (2004)
5. Gomma, H.: A software design method for real-time systems. Commun ACM **27**(9), 938–949 (1984)
6. Hatley, D., Pirabhai, I.: Strategies for Real-time System Specificaiton. Dorset House, New York (1988)
7. Ward, P., Mellor, S.: Structured Development for Real-time Systems. Prentice Hall, Upper Saddle River, NJ (1985)
8. Parnas, D., Clements, P., Weiss, D.: The modular structure of complex systems. In: IEEE Conference on, Software Engineering, pp. 551–556 (1984)
9. Wang, S., Merrick, J.R., Shin, K.G.: Component allocation with multiple resource constraints for large embedded real-time software design. In: IEEE Real-Time and Embedded Technology and Applications, Symposium, pp. 219–226 (2004)
10. Wang, S., Shin, K.G.: Task construction for model-based design of embedded control software. IEEE Trans. Softw. Eng. **32**(4), 254–264 (2006)
11. Davare, A., Zhu, Q., Natale, M.D., Pinello, C., Kanajan, S., Sangiovanni-vincentelli, A.L.: Period optimization for hard real-time distributed automotive systems. In: Design Automation Conference, pp. 278–283 (2007). <error l="82" c="Undefined command" />10.1109/DAC.2007.375172.

12. Natale, M.D., Zheng, W., Pinello, C., Giusto, P., Sangiovanni-Vincentelli, A.L.: Optimizing end-to-end latencies by adaptation of the activation events in distributed automotive systems. In: IEEE Real-Time and Embedded Technology and Applications Symposium, (2007)
13. Alur, R., Etessami, K., Torre, S.L., Peled, D.: Parametric temporal logic for "Model Measuring". ACM Trans. Comput. Logic **2**(3), 388–407 (2001)
14. Dixit, M.G., Dasgupta, P., Ramesh, S.: Taming the component timing: A CBD methodology for component based embedded systems. In: Design Automation and Test in Europe (DATE), (2009)
15. Dixit, M., Ramesh, S., Dasgupta, P.: Some results on parametric temporal logic. Inf. Process. Lett. **111**(3), 994–998 (2011)
16. Dixit, M.G.: Formal methods for early time-budgeting in component based embedded control systems. Ph.D. thesis, IIT Kharagpur (2012)
17. Bartolini, C., Lipari, G., Natale, M.D.: From functional blocks to the synthesis of the architectural model in embedded real-time applications. In: IEEE Real-Time and Embedded Technology and Applications, Symposium, pp. 458–467 (2005)
18. Hamann, A., Jersak, M., Richter, K., Ernst, R.: Design space exploration and system optimization with SymTA/S - symbolic timing analysis for systems. In: IEEE Real-Time Systems, Symposium, pp. 469–478 (2004)
19. AUTOSAR Consortium: AUTomotive Open System ARchitecture. http://www.autosar.org
20. Klobedanz, K., Kuznik, C., Thuy, A., Mueller, W.: Timing modeling and analysis for AUTOSAR-based software development - A case study. In: Design Automation and Test in, Europe, pp. 642–645 (2010)
21. Alur, R., Henzinger, T.: Real-time logics: Complexity and expressiveness. Inf. Comput. **104**(1), 35–77 (1993)
22. Clarke, E.M., Grumberg, O., Peled, D.A.: Model Checking. The MIT Press, Cambridge, MA, USA (2000)
23. Emerson, A.E., Mok, A., Sistla, A., Srinivasan, J.: Quantitative temporal reasoning. In: Computer Aided Verification, pp. 136–145 (1994)
24. Emerson, A.E., Trefler, R.: Parametric quantitative temporal reasoning. In: IEEE Symposium on Logic in Computer Science, pp. 336–343 (1999)
25. Pnueli, A.: The temporal logic of programs. In: 18th IEEE. Foundations of Computer Science, pp. 46–57 (1977)
26. Eclipse Modeling Framework. http://www.eclipse.org
27. NuSMV: A New Symbolic Model Checker. http://nusmv.fbk.eu//
28. Yices: An SMT Solver. http://yices.csl.sri.com/

Chapter 8
Contract-Based Reasoning for Component Systems with Rich Interactions

Susanne Graf, Roberto Passerone and Sophie Quinton

Abstract In this chapter we propose a rule unifying circular and non-circular assume-guarantee reasoning and show its interest for contract-based design and verification. Our work was motivated by the need to combine, in the top-down methodology of the FP7 SPEEDS project, partial tool chains for two component frameworks derived from the HRC model and using different refinement relations. While the L0 framework is based on a simple trace-based representation of behaviors and uses set operations for defining refinement, the more elaborated L1 framework offers the possibility to build systems of components with complex interactions. Our approach in L1 is based on circular reasoning and results in a method for checking contract dominance which does not require the explicit composition of contracts. In order to formally relate results obtained in L0 and L1, we provide a definition of the minimal concepts required by a consistent contract theory and propose abstract definitions which smoothly encompass hierarchical components. Finally, using our relaxed rule for circular reasoning, we show how to use together the L0 and L1 refinement relations and as a result their respective tool chains.

S. Graf (✉)
VERIMAG/CNRS, 2 avenue de Vignate, 38610 Gières, France
e-mail: susanne.graf@imag.fr

R. Passerone
DISI/University of Trento,via Sommarive 5, 38123 Trento, Italy
e-mail: roberto.passerone@unitn.it

S. Quinton
IDA/TU Braunschweig, Hans-Sommer-Straße 66, 38106 Braunschweig, Germany
e-mail: quinton@ida.ing.tu-bs.de

A. Sangiovanni-Vincentelli et al. (eds.), *Embedded Systems Development*, Embedded Systems 20, DOI: 10.1007/978-1-4614-3879-3_8,

8.1 Introduction

Contract and interface frameworks are emerging as the formalism of choice for system designs that require large and distributed teams, or where the supply chain is complex [1–3]. This style of specification is typically employed for top-down design of systems of components, where the system under design is built by a sequence of decomposition and verification steps. In this chapter we present and study some distinctive features of contract theories for frameworks in which the interaction between components is "rich", i.e., more complex than the usual input/output (I/O) communication. One such component framework is BIP [3] which allows multi-party synchronizations scheduled according to priorities. In addition, we show how to combine results obtained using different contract refinement relations.

Our work has its practical motivation in the component framework HRC [2–5] (standing for Heterogeneous Rich Components) defined in the FP7 IP project SPEEDS [6], which has been reused in the FP7 STREP project COMBEST [7] and the ARTEMIS project CESAR [8]. The HRC model defines component properties in terms of extended transition systems and provides several composition models, ranging from low-level semantic composition to composition frameworks underlying the design tools used by system designers. More precisely, HRC is organized around two abstraction levels called L0 and L1 and describing respectively the *core level* and the *analysis tool level* of HRC [9]. That is, L0 determines the expressive power of the entire model and there exist translations from L1 models to L0. On the other hand, L1 extends the core model with concepts such as coordination mechanisms — the rich interactions mentioned in the title. Analysis tools can then take advantage of these additional concepts to make system descriptions more concise and therefore verification more efficient.

Our objective is to allow combined use of synchronous tools like Simulink [10] for L0 and synchronization-based tools like BIP for L1, which have complementary strengths. For example, Simulink is very convenient for modeling physical dynamic systems or streaming applications. In contrast BIP, which encompasses rich interactions, is well adapted for describing the dynamic behavior of sets of components depending on available resources for memory, energy, communication bandwidth etc. We are interested in this chapter in the relation between the L0 and L1 contract frameworks as we want to use verification results established in L1 for further reasoning within L0. The presence of rich interactions in L1 makes contract composition problematic and leads us to focus instead on *circular reasoning*, which allows a component and its environment to be refined concurrently—each one relying on the abstract description of its context—and entails an interesting rule for proving *dominance*, i.e., refinement between contracts. In order to relate L0 and L1, we define a generic contract framework that uses abstract composition operators and thus encompasses a variety of interaction models, including those for L0 and L1. Finally, we show how to use a relaxed rule for circular reasoning to combine partial tool chains for both frameworks into a complete tool chain for our methodology.

To the best of our knowledge, this is the first time that a rule combining different refinement relations is proposed and used to unify two contract frameworks. While circular reasoning has been extensively studied, e.g., in [11, 12], existing work focuses on finding sufficient conditions for soundness of circular reasoning while we focus on how to use circular reasoning in a contract-based methodology. Non-circular assume-guarantee reasoning is also a topic of intense research focused on finding a decomposition of the system that satisfies the strong condition imposed on at least one of its components [13]. Finally, our contract frameworks are related to interface automata [14]. Since de Alfaro and Henzinger's seminal paper many contract and interface theories have been developed for numerous frameworks (see e.g. [15–20] to name just a few). However these theories focus on composition of contracts while we strive to avoid that and furthermore they do not handle rich interactions. Examples include [20, 21] based on modal I/O automata and [16] defining relational interfaces for capturing functional dependencies between inputs and outputs of an interface. Preliminary versions of our contract framework appeared in [22, 23] but did not address the question of combining results obtained for different refinements.

This chapter is structured as follows: Sect. 8.2 describes our design and verification methodology as well as generic definitions of component and contract framework. It then discusses sufficient reasoning rules for establishing dominance without composing contracts. Section 8.3 presents how the proposed approach is applied to the L0 and L1 frameworks. In particular it shows how their different satisfaction relations may be used together using *relaxed circular reasoning* and discusses practical consequences of this result. Section 8.4 concludes the chapter. The proofs of all theorems presented in this paper are described in [24].

8.2 Design Methodology

Our methodology is based on an abstract notion of component. We characterize a component K by its *interface* defined as a set $\mathscr{P}$ of *ports* which describe what can be observed by its environment. We suppose given a global set of ports $Ports$, which all sets of ports in the following are subsets of. In addition, components are also characterized by their *behavior* . At this level of abstraction, we are not concerned with how behaviors are represented and develop our methodology independently of the particular formalism employed. Interactions (potentially complex) between components are expressed using the concept of *glue* operator [25] . A glue defines how the ports of different components are connected and the kind of synchronization and data exchange that may take place. We denote the composition of two components K_1 and K_2 through a glue gl as $gl\{K_1, K_2\}$. The glue must be defined on the union of the ports $\mathscr{P}_1$ and $\mathscr{P}_2$ of the components.

In order to separate the implementation phase of a component from its integration into the system under design, we use *contracts* [5, 22, 26] . A contract for a component K describes the interface $\mathscr{P}$ of K, the interaction between K and its environment E, the expected behavior of E, called the *assumption* A of the contract, and the expected

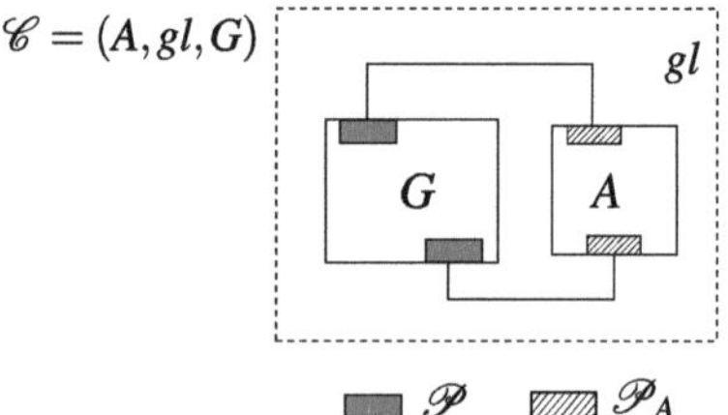

Fig. 8.1 A contract (A, gl, G) for an interface $\mathscr{P}$

behavior of K, called the *guarantee* G. Assumptions and guarantees are in turn expressed as components, defining the interface and the behavior that are considered acceptable from the environment and from the component. Thus, formally, a contract $\mathscr{C}$ for an interface $\mathscr{P}$ is a triple (A, gl, G), where gl is a glue operator on $\mathscr{P} \cup \mathscr{P}_A$ for some $\mathscr{P}_A$ disjoint from $\mathscr{P}$; the assumption A is a component with interface $\mathscr{P}_A$; and the guarantee G is a component with interface $\mathscr{P}$. Note that the interface of the environment is implicitly defined by gl. Graphically, we represent contracts as in Fig. 8.1.

From a macroscopic point of view, we adopt a top-down design and verification methodology (see Fig. 8.2) in which global requirements are pushed progressively from the top-level system to the low-level atomic components. As usual, this is just a convenient representation; in real life, the final picture is always obtained in several iterations alternatively going up and down the hierarchy [27].

While the refinement relation between a specification and an implementation is at the core of component-based design, in contract-based design refinement takes different forms depending on whether it relates a system to a specification, two contracts or an implementation to a contract. In this chapter we use a methodology which divides the design and verification process into three steps corresponding to these three forms of refinement.

We assume that the system K under construction has to realize a global requirement φ together with an environment on which we may have some knowledge, expressed by a property A. Both φ and A are expressed w.r.t. the interface $\mathscr{P}$ of K. We proceed as follows: (1) define a *contract* $\mathscr{C} = (A, gl, G)$ for $\mathscr{P}$ such that $gl\{A, G\}$ *conforms* to φ; (2) decompose K as subcomponents K_i connected through a glue operator gl_I and provide a contract $\mathscr{C}_i$ for each of them; possibly iterate this step if needed; (3) prove that whenever a set of implementations K_i *satisfy* their contracts $\mathscr{C}_i$, then their composition satisfies the top-level contract $\mathscr{C}$ (*dominance*) — and thus guarantee φ; (4) provide such implementations.

The correctness proof for a particular system is therefore split into 3 phases: *conformance* (denoted $\preccurlyeq$) of the system defined by the top-level contract $\mathscr{C}$ to φ; *dominance* of $\mathscr{C}$ by the composition of the set of contracts $\{\mathscr{C}_i\}$ through gl_I; and *satisfaction* (denoted $\models$) of each $\mathscr{C}_i$ by the corresponding implementation K_i. Thus, *conformance* relates closed systems, *dominance* relates contracts, while *satisfaction* relates components to contracts.

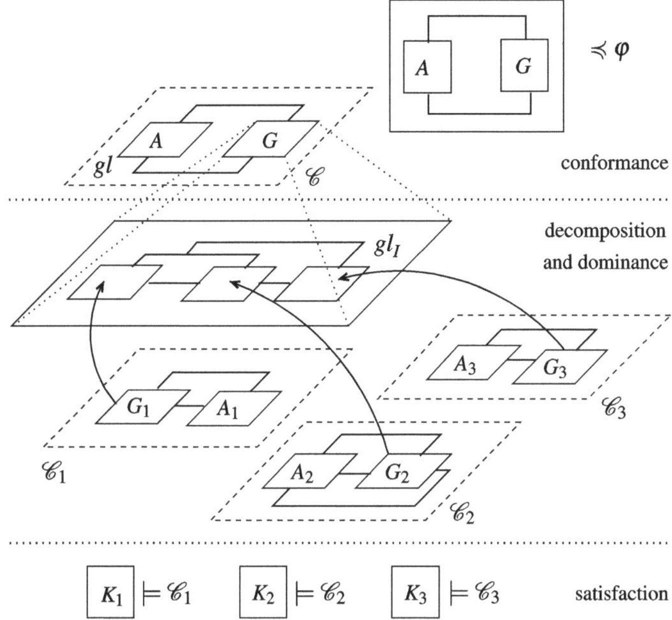

Fig. 8.2 Proof of $gl\{A, gl_I\{K_1, K_2, K_3\}\} \preccurlyeq \varphi$

Note that the assumption of $\mathscr{C}_1$ is represented as one component A_1 while in the actual system K_1 will be used in the context of three components, namely K_2, K_3 and A. Thus, we need to relate the actual glues gl and gl_I to the glue gl_1 of $\mathscr{C}_1$. In other words, we need a glue gl_{E_1} to compose K_2, K_3 and A as well as an operation $\circ$ on glues such that $gl \circ gl_I = gl_1 \circ gl_{E_1}$. In most cases, $\circ$ cannot simply be composition of functions and has to involve some *flattening* of the system.

8.2.1 Contract Framework

To summarize, we consider a component framework that smoothly supports complex composition operators and hierarchical components. The elements of the component framework are as follows:

Definition 1 *(Component framework).* A *component framework* is defined by a tuple $(\mathscr{K}, GL, \circ, \cong)$ where:

- $\mathscr{K}$ is a set of *components*. Each component $K \in \mathscr{K}$ has as *interface* a set of *ports*, denoted $\mathscr{P}_K$ and subset of our global set of ports $Ports$.

- GL is a set of *glues*. A glue is a partial function $2^{\mathscr{K}} \longrightarrow \mathscr{K}$ transforming a set of components into a new composite component. Each $gl \in GL$ is defined on a set of ports S_{gl}, called *support set*, and defines a new interface $\mathscr{P}_{gl}$ for the new component, called *exported interface*. $K = gl(\{K_1, \ldots, K_n\})$ is defined if $K_1, \ldots, K_n \in \mathscr{K}$ have disjoint interfaces, $S_{gl} = \bigcup_{i=1}^{n} \mathscr{P}_{K_i}$ and $\mathscr{P}_K = \mathscr{P}_{gl}$.
- $\circ$ is a partial operator on GL, called *flattening*, to compose glues. $gl \circ gl'$ is defined if $\mathscr{P}_{gl'} \subseteq S_{gl}$. Its support set is $S_{gl} \backslash \mathscr{P}_{gl'} \cup S_{gl'}$ and its interface is $\mathscr{P}_{gl}$.
- $\cong \ \subseteq \mathscr{K} \times \mathscr{K}$ is an equivalence relation between components.

We simplify our notation by writing $gl\{K_1, \ldots, K_n\}$ instead of $gl(\{K_1, \ldots, K_n\})$. The equivalence relation $\cong$ is typically used for relating composite components with their semantics given as an atomic component. More importantly, $\circ$ must be coherent with $\cong$ in the sense that $gl\{gl'\{\mathscr{K}_1\}, \mathscr{K}_2\} \cong (gl \circ gl')\{\mathscr{K}_1 \cup \mathscr{K}_2\}$ for any sets of components $\mathscr{K}_i$ such that all terms are defined.

After formalizing generic properties required from a component framework, we now define the relations used in the methodology for dealing with contracts. Satisfaction is usually considered as a derived relation and chosen as the weakest relation implying conformance and preserved by composition. We loosen the coupling between satisfaction and conformance to obtain later stronger reasoning schemata for dominance. Furthermore, we propose a representation of satisfaction as a set of *refinement under context* relations denoted $\sqsubseteq_{A,gl}$ and such that $K \sqsubseteq_{A,gl} G$ iff $K \models (A, gl, G)$.

Definition 2 *(Contract framework)* A *contract framework* is defined by a tuple $(\mathscr{K}, GL, \circ, \cong, \preccurlyeq, \models)$ where:

- $(\mathscr{K}, GL, \circ, \cong)$ is a component framework.
- $\preccurlyeq \ \subseteq \mathscr{K} \times \mathscr{K}$ is a preorder called *conformance* relating components having the same interface.
- $\models$ is a relation called *satisfaction* between components and contracts s.t.: the relations $\sqsubseteq_{A,gl}$ defined by $K \sqsubseteq_{A,gl} G$ iff $K \models (A, gl, G)$ are preorders; and, if $K \models (A, gl, G)$ then $gl\{A, K\} \preccurlyeq gl\{A, G\}$.

Our definition of satisfaction emphasizes the fact that $\models$ can be seen as a set of refinement relations where $K \sqsubseteq_{A,gl} G$ means that K refines G in the context of A and gl. The condition which relates satisfaction and conformance ensures that the actual system $gl\{A, K\}$ will conform to the global requirement φ discussed in the methodology because $\preccurlyeq$ is transitive and $gl\{A, G\} \preccurlyeq \varphi$.

Example 1 Typical notions of conformance for labeled transition systems are *trace inclusion* and its structural counterpart *simulation*. For these, satisfaction is usually defined as the weakest relation implying conformance.

$$K \models (A, gl, G) \triangleq gl\{K, A\} \preccurlyeq gl\{G, A\}$$

Dominance is a key notion for reasoning about contracts rather than using refinement between components. Proving that a contract $\mathscr{C}$ dominates $\mathscr{C}'$ means showing that every component satisfying $\mathscr{C}$ also satisfies $\mathscr{C}'$.[1] However, a dominance check involves in general not just a pair of contracts: a typical situation would be the one depicted in Fig. 8.2, where a set of contracts $\{\mathscr{C}_i\}_{i=1}^n$ are attached to disjoint interfaces $\{\mathscr{P}_i\}_{i=1}^n$. Besides, a glue gl_I is defined on $P = \bigcup_{i=1}^n \mathscr{P}_i$ and a contract $\mathscr{C}$ is given for P. In this context, a set of contracts $\{\mathscr{C}_i\}_{i=1}^n$ dominates a contract $\mathscr{C}$ w.r.t. a glue gl_I if any set of components satisfying contracts $\mathscr{C}_i$, when composed using gl_I, makes a component satisfying $\mathscr{C}$.

Definition 3 *(Dominance)* Let $\mathscr{C}$ be a contract on $\mathscr{P}$, $\{\mathscr{C}_i\}_{i=1}^n$ a set of contracts on $\mathscr{P}_i$ and gl_I a glue such that $S_{gl_I} = \bigcup_{i=1}^n \mathscr{P}_i$ and $\mathscr{P} = \mathscr{P}_{gl_I}$. Then $\{\mathscr{C}_i\}_{i=1}^n$ *dominates* $\mathscr{C}$ *with respect to* gl_I iff for all components $\{K_i\}_{i=1}^n$:

$$(\forall i : K_i \models \mathscr{C}_i) \implies gl_I\{K_1, \ldots, K_n\} \models \mathscr{C}$$

Note that this formal definition of dominance does not help establishing dominance in practice because looking at all possible components satisfying a contract is not realistic. What we need is a sufficient condition that refers to assumptions and guarantees, rather than components. One such condition exists when the composition of the low-level guarantees G_i satisfies the top-level contract $\mathscr{C}$ and furthermore each low-level assumption A_i is discharged by the abstraction of its environment defined by the guarantees of the other components. Formally:

$$\begin{cases} gl_I\{G_1, \ldots, G_n\} \models \mathscr{C} \\ \forall i : gl_{E_i}\{A, G_1, \ldots, G_{i-1}, G_{i+1}, \ldots, G_n\} \models \mathscr{C}_i^{-1} \end{cases} \tag{8.1}$$

where for any contract $\mathscr{C}_i = (A_i, gl_i, G_i)$ we use the notation $\mathscr{C}_i^{-1}$ to denote the contract (G_i, gl_i, A_i).

In the next subsection, we provide two rules which indeed make the previous condition sufficient for establishing dominance: one is similar to circular assume-guarantee reasoning and the other one deals with preservation of satisfaction by composition. This result is particularly significant because one can check dominance while avoiding composition of contracts, which is impossible in the general case and leads to state explosion in most concrete contract frameworks.

[1] One may also need to ensure that the assumptions of the low-level contracts are indeed satisfied in the actual system. This is achieved by strengthening the definition with:

$$\forall E \text{ on } \mathscr{P}_A, \text{ if } E \models (G', gl', A') \text{ then } E \models (G, gl, A)$$

8.2.2 Reasoning Within a Contract Framework

We use here the representation of satisfaction as a set of refinement under context relations $\sqsubseteq_{A,gl}$ where $K \sqsubseteq_{A,gl} G$ if and only if $K \models (A, gl, G)$. The usual non-circular assume-guarantee rule reads as follows in our context:

$$K \sqsubseteq_{A,gl} G \wedge E \sqsubseteq A \implies K \sqsubseteq_{E,gl} G \tag{8.2}$$

where $E \sqsubseteq A$ denotes that for any component G and gl such that $\sqsubseteq_{G,gl}$ is defined $E \sqsubseteq_{G,gl} A$. This rule relates the behavior of K, when composed with the abstract environment A, to the behavior of K, when composed with its actual environment E. However it is quite limited as it imposes a very strong condition on E. Hence the following rule which is commonly referred to as *circular reasoning*.

$$K \sqsubseteq_{A,gl} G \wedge E \sqsubseteq_{G,gl} A \implies K \sqsubseteq_{E,gl} G$$

Note that E and K may symmetrically rely on each other. For a given contract framework, this property can be proven by an induction based on the semantics of composition and refinement. Unfortunately, circular reasoning is not sound in general. In particular it does not hold for parallel composition with synchronizations (as in Petri nets or process algebras) or instantaneous mutual dependencies between inputs and outputs (as in synchronous formalisms). The following example illustrates one possible reason for the non validity of circular reasoning.[2]

Example 2 Consider a contract framework where components are labeled transition systems and composition is strong synchronization between corresponding labels and interleaving of others denoted $\|$. Define conformance as simulation and satisfaction as the usual relation defined in Example 1. The circular reasoning rule translates into: if $K \| A$ is simulated by $G \| A$ and $E \| G$ is simulated by $A \| G$ then $K \| E$ is simulated by $G \| E$. In the example of Fig. 8.3, both G and A forbid a synchronization between b_K and b_E from occurring. This allows their respective refinements according to $\sqsubseteq^{\preccurlyeq}$, namely K and E, to offer respectively b_K and b_E, since they can rely on respectively G and A to forbid their actual occurrence. But obviously, the composition $K \| E$ now allows a synchronization between b_K and b_E.

Note that this satisfaction relation can be strengthened to obtain a more restrictive relation for which circular reasoning is sound. This is the approach taken for the L1 contract framework in Sect. 8.3.2, where we need circular reasoning to avoid composition of contracts.

A second rule which is used for compositional reasoning in most frameworks is: if $I \sqsubseteq S$, then $I \| E \sqsubseteq S \| E$. It states that if an implementation I refines its specification S then it refines it in any environment E. The equivalent of this rule for satisfaction is more complex as refinement here relates closed systems.

---

[2] Note that non-determinism is another reason here for the non validity of circular reasoning.

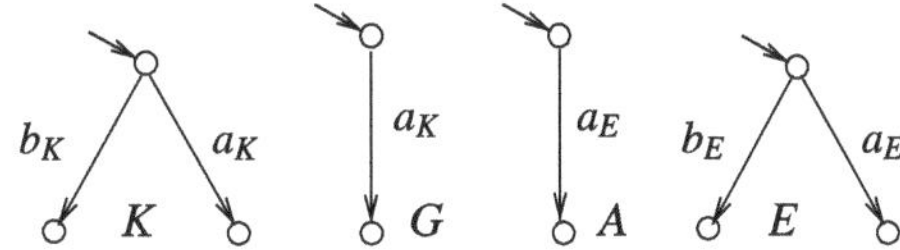

Fig. 8.3 $K \parallel A \preccurlyeq G \parallel A$ and $E \parallel G \preccurlyeq A \parallel G$ but $K \parallel E \not\preccurlyeq G \parallel E$

Definition 4 Satisfaction $\models$ is preserved by composition iff for any component E, gl such that $S_{gl} = \mathscr{P}_E \cup \mathscr{P}$ for some $\mathscr{P}$ such that $\mathscr{P} \cap \mathscr{P}_E = \emptyset$ and gl_E, E_1, E_2 such that $E = gl_E\{E_1, E_2\}$, the following holds for any components I, S on $\mathscr{P}$:

$$I \sqsubseteq_{E,gl} S \implies gl_1\{I, E_1\} \sqsubseteq_{E_2,gl_2} gl_1\{S, E_1\}$$

where gl_1 and gl_2 are such that $gl \circ gl_E = gl_2 \circ gl_1$.

We now have the ingredients to formalize our sufficient condition for dominance. This condition reduces a dominance proof to a set of satisfaction checks, one for the refinement between the guarantees and n for discharging individual assumptions.

Theorem 1 *Suppose that circular reasoning is sound and satisfaction is preserved by composition. If* $\forall i \; \exists gl_{E_i} : gl \circ gl_I = gl_i \circ gl_{E_i}$ *then to prove that* $\{\mathscr{C}_i\}_{i=1}^n$ *dominates* $\mathscr{C}$ *w.r.t. gl, it is sufficient to prove that condition* (1) *holds.*

8.3 Circular Reasoning in Practice

In this section, we show how the results presented in the previous section have been applied within the SPEEDS project: we define two contract frameworks, called L0 and L1, and show how to combine them.

8.3.1 The L0 Framework

A component K with interface $\mathscr{P}_K$ at level L0 of HRC is defined as a set of behaviors in the form of traces, or runs, over $\mathscr{P}_K$. The behaviors correspond to the history of values seen at the ports of the component for each particular behavior. For instance, these histories could be the traces generated by a labeled transition system (LTS). Composition is defined as a composite that retains only the matching behaviors of the components. If the ports of the two components have the same names, composition at the level of trace sets boils down to a simple intersection of the sets of behaviors. Because in our framework components must have disjoint sets of ports under composition, we must introduce glues, or connectors, as explicit components that establish a synchronous relation between the histories of connected ports. The collection of these simple connectors forms the glues $gl \in GL$ of our framework at the L0 level.

We can model a glue as an extra component K_{gl}, whose set of ports includes all the ports of the components involved in the composition. This component has as the set of behaviors all the identity traces. Composition can then be taken as the intersection of the sets of behaviors of the components, together with the glue. To make this work, we must also equalize the ports of all trace sets using inverse projection $\mathbf{proj}^{-1}_{\mathscr{P}_i,\mathscr{P}}$, which extends behaviors over $\mathscr{P}_1$ with the appropriate additional ports of $\mathscr{P}$. If we denote the interface of the composite as $\mathscr{P}_{gl}$, and if $\mathscr{K} = \{K_1, \dots, K_n\}$ is a set of components such that $\mathscr{P}_1, \dots, \mathscr{P}_n$ are pairwise disjoint, then a glue gl for $\mathscr{K}$ is a component K_{gl} defined on the ports $\mathscr{P} = \mathscr{P}_{gl} \cup (\bigcup_{i=1}^{n} \mathscr{P}_i)$, and:

$$\begin{aligned} K &= gl\{K_1, \dots, K_n\} \\ &= \mathbf{proj}_{P_{gl},\mathscr{P}} \left(K_{gl} \cap \mathbf{proj}^{-1}_{\mathscr{P}_1,\mathscr{P}} (K_1) \cap \dots \cap \mathbf{proj}^{-1}_{\mathscr{P}_n,\mathscr{P}} (K_n) \right) \end{aligned}$$

The definition of $\circ$ is straightforward: since glues are themselves components, their composition follows the same principle as component composition. Finally, the $\cong$ relation on $\mathscr{K}$ is taken as equality of sets of traces.

In the L0 model there exists a unique maximal component satisfying a contract $\mathscr{C}$, namely $M_{\mathscr{C}} = G \cup \neg A$, where $\neg$ denotes the operation of complementation on the set of all behaviors over ports $\mathscr{P}_A$. A contract $\mathscr{C} = (A, G)$ is in *canonical form* when $G = M_{\mathscr{C}}$. Every contract has an equivalent contract in canonical form, which is obtained by replacing G with $M_{\mathscr{C}}$. The operation of computing a canonical form is well defined, since the maximal implementation is unique, and it is idempotent. It is easy to show that $K \models \mathscr{C}$ if and only if $K \subseteq M_{\mathscr{C}}$.

The L0 contract framework has strong compositional properties, which derive from its simple definition and operators [26]. The theory, however, depends on the effectiveness of certain operators, complementation in particular, which are necessary for the computation of canonical forms. While the complete theory can be formulated without the use of canonical forms, complementation remains fundamental in the definition of contract composition, which is at the basis of system construction. Circular reasoning is not sound for contracts which are *not* in canonical form (Example 2 is a counter-example in that case). This is a limitation of the L0 framework, since working with canonical forms could prove computationally hard.

8.3.2 The L1 Framework

L1 composition is based on *interactions*, which involve non-empty sets of ports. An interaction is defined by the components that synchronize when it takes place and the ports through which these components synchronize. Interactions are structured into connectors which are used as a mechanism for *encapsulation*: only these connectors appear at the interface of a composite component. This enables to abstract the behavior of a component in a black-box manner, by describing which connector is triggered but not exactly which interaction takes place. Furthermore L1 is expressive enough to encompass synchronous systems.

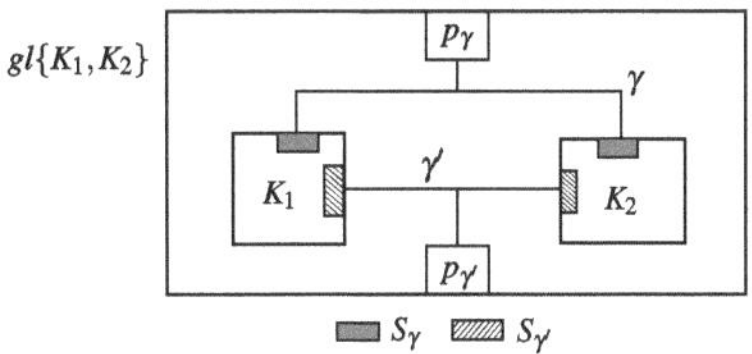

Fig. 8.4 The role of connectors in a composition

Definition 5 An *atomic component* on an interface $\mathscr{P}$ is defined by an LTS $K = (Q, q^0, 2^{\mathscr{P}}, \longrightarrow)$, where Q is a set of states, q^0 an initial state and $\longrightarrow \subseteq Q \times 2^{\mathscr{P}} \times Q$ is a transition relation.

Note that atomic components are labeled by sets of ports rather than ports because we allow several ports of a component to be triggered at the same time.

Definition 6 An *interaction* is a non-empty set of ports. A *connector* γ is defined by a set of ports S_γ called the *support set* of γ, a port p_γ called its *exported port* and a set $\mathfrak{I}(\gamma)$ of interactions in S_γ.

The notions of support set and exported port are illustrated in Fig. 8.4, where connectors relate in a composition a set of inner ports (of the subcomponents) to an outer port (of the composite component). One should keep in mind that a connector γ, and thus the exported port p_γ, represents a set of interactions rather than a single interaction.

Typical connectors represent rendezvous (only one interaction, equal to the support set), broadcast (all the interactions containing a specific port called *trigger*) and also mutual exclusion (some interactions but not their union).

We now define glues as sets of connectors which may be used together in order to compose components.

Definition 7 A glue gl on a support set S_{gl} is a set of connectors with distinct exported ports and with support sets included in S_{gl}.

A glue gl defines as exported interface $\mathscr{P}_{gl}$ the set $\{p_\gamma \mid \gamma \in gl\}$. Besides, $\mathfrak{I}(gl)$ denotes the set of all interactions of the connectors in gl, i.e.: $\mathfrak{I}(gl) = \bigcup_{\gamma \in gl} \mathfrak{I}(\gamma)$. In Fig. 8.4, gl is composed of connectors γ and γ' and defines a composite component denoted $gl\{K_1, K_2\}$.

Definition 8 A *component* is either an atomic component or it is inductively defined as the composition of a set of components $\{K_i\}_{i=1}^n$ with disjoint interfaces $\{\mathscr{P}_i\}_{i=1}^n$ using a glue gl on $\mathscr{P} = \bigcup_{i=1}^n \mathscr{P}_i$. Such a composition is called a *composite component* on $\mathscr{P}_{gl}$ and it is denoted $gl\{K_i\}_{i=1}^n$.

So far, we have defined components and glues. Glues can be composed so as to allow flattening of components. Such a composition requires to handle *hierarchical connectors* built by merging connectors defined at different levels of hierarchy. The definition of the operator $\circ$ used for this purpose is omitted here and can be found in [24]. Connectors whose exported ports and support sets are not related are called *disjoint* and need not be composed. The operator $\circ$ is then easily extended to glues:

the composition $gl \circ gl'$ of two glues gl and gl' is obtained from $gl \cup gl'$ by inductively composing all connectors which are not disjoint.

We can now formally define the flattened form of a component. This in turn will allow us to provide an equivalence relation between components based on the semantics of their flattened form. A component is called *flat* if it is atomic or of the form $gl\{K_1, \ldots, K_n\}$, where all K_i are atomic components. A component that is not flat is called *hierarchical*. A hierarchical component K is of the form $gl\{K_1, \ldots, K_n\}$ such that at least one K_i is composite. Thus, such a K can be represented as $gl\{gl'\{\mathscr{K}^1\}, \mathscr{K}^2\}$, where $\mathscr{K}^1$ and $\mathscr{K}^2$ are sets of components.

Definition 9 The *flattened form* of a component K is denoted $\flat(K)$ and defined inductively as:

- if K is a flat component, then $\flat(K)$ is equal to K.
- otherwise, K is of the form $gl\{gl'\{\mathscr{K}^1\}, \mathscr{K}^2\}$, and then $\flat(K)$ is the flattened form of $(gl \circ gl')\{\mathscr{K}^1 \cup \mathscr{K}^2\}$.

Definition 10 The *semantics* $[\![K]\!]$ of a flat component $K = gl\{K_1, \ldots, K_n\}$ is defined as $(Q, q^0, \mathfrak{I}(gl), \longrightarrow)$, where $Q = \prod_{i=1}^n Q_i$, $q^0 = (q_1^0, \ldots, q_n^0)$ and $\longrightarrow$ is such that: given two states $q^1 = (q_1^1, \ldots, q_n^1)$ and $q^2 = (q_1^2, \ldots, q_n^2)$ in Q and an interaction $\alpha \in \mathfrak{I}(gl)$, $q^1 \xrightarrow{\alpha} q^2$ if and only if $\forall i, q_i^1 \xrightarrow{\alpha_i}_i q_i^2$, where $\alpha_i = \alpha \cap \mathscr{P}_i$.

We use the convention that $\forall q : q \xrightarrow{\emptyset} q$ so components not involved in an interaction do not move. Thus the semantics of a flat component is obtained as the composition of its constituting LTS where labels are synchronized according to the interactions of $\mathfrak{I}(gl)$.

We then define equivalence $\cong$ as follows: two components are equivalent if their flattened forms have the same semantics. Note that in practice one would prefer to define the semantics of a hierarchical component as a function of the semantics of its constituting components. In presence of encapsulation this requires to distinguish between closed and open systems and thus to provide two different semantics. Details can be found in [24].

We now have the ingredients for defining the L1 component framework and we focus on its contract framework.

Definition 11 $K_1 \preccurlyeq^{L1} K_2$ if and only if $[\![K_1]\!]$ is simulated by $[\![K_2]\!]$.

Thus L1-conformance is identical to L0-conformance for components without non-observable non-determinism, and otherwise stronger. Note that in verification tools, in order to check trace inclusion efficiently, one will generally check simulation anyway. Satisfaction is defined as follows.

Definition 12 A component K *satisfies* a contract $\mathscr{C} = (A, gl, G)$ for $\mathscr{P}_K$, denoted $K \models^{L1} (A, gl, G)$, if and only if:

$$\begin{cases} gl\{K, A_{det}\} \preccurlyeq^{L1} gl\{G, A_{det}\} \\ (q_K, q_A)\, \mathfrak{R}\, (q_G, q'_A) \wedge \exists q'_K : q_K \xrightarrow{\alpha}_K q'_K \implies \exists q'_G : q_G \xrightarrow{\alpha}_G q'_G \end{cases}$$

where A_{det} is the determinization of A, $\mathcal{R}$ is the relation on states proving that $gl\{K, A_{det}\} \preccurlyeq^{L1} gl\{G, A_{det}\}$ and $\alpha \in 2^{\mathcal{P}_K}$ is such that $\exists \alpha' \in \mathfrak{I}(gl) : \alpha \subseteq \alpha'$.

Thus $\models^{L1}$ strengthens the satisfaction relation used in the L0 framework by: 1) determinizing A; 2) requiring every transition of K to have a counterpart in each related state of G — unless it is structurally forbidden by gl — but the target states of the transition need to be related only if the environment allows this transition. As a consequence, $\models^{L1}$ allows circular reasoning.

8.3.3 Relaxed Circular Reasoning

We have presented in the previous sections two contract frameworks developed in the SPEEDS project. We show now how we use their respective tool chains together. Unifying the L0 and L1 component frameworks is quite straightforward. Nevertheless, we have introduced two different notions of satisfaction: $\models^{L0}$ and $\models^{L1}$ where the second one is strictly stronger than the first one. To combine results based on L0 and L1, we propose a rule called *relaxed circular reasoning* for two (possibly different) refinement relations:

$$K \sqsubseteq^1_{A,gl} G \wedge E \sqsubseteq^2_{G,gl} A \implies K \sqsubseteq^1_{E,gl} G \tag{8.3}$$

This rule generalizes circular and non-circular reasoning by not restricting $\sqsubseteq^2_{G,gl}$ to refinement under context $\sqsubseteq^1_{G,gl}$ or refinement in any context $\sqsubseteq^1$. Depending on which relation is the most restrictive it can be used in two different ways:

1. If the first relation allows circular reasoning and is stronger than the second one (i.e. $K \sqsubseteq^1_{A,gl} G \implies K \sqsubseteq^2_{A,gl} G$) then our new rule relaxes circular reasoning by requiring $E \sqsubseteq^2_{G,gl} A$ rather than $E \sqsubseteq^1_{G,gl} A$.
2. Symmetrically, if the first relation does not allow circular reasoning and refinement in any context $\sqsubseteq^1$ is stronger than the second one then this rule relaxes non circular reasoning by requiring $E \sqsubseteq^2_{G,gl} A$ rather than $E \sqsubseteq^1 A$.

Interestingly, relaxed circular reasoning can be used both ways for L0- and L1-satisfaction. First it leads to a relaxed sufficient condition for dominance in L1.

Theorem 2 $K \sqsubseteq^{L1}_{A,gl} G \wedge E \sqsubseteq^{L0}_{G,gl} A$ *implies* $K \sqsubseteq^{L1}_{E,gl} G$.

Theorem 3 *If* $\forall i\ \exists gl_{E_i} : gl \circ gl_I = gl_i \circ gl_{E_i}$ *the following is sufficient to prove that* $\mathcal{C}$ *dominates* $\{\mathcal{C}_i\}_{i=1..n}$ *w.r.t.* gl*:*

$$\begin{cases} gl_I\{G_1, \ldots, G_n\} \models^{L1} \mathcal{C} \\ \forall i : gl_{E_i}\{A, G_1, \ldots, G_{i-1}, G_{i+1}, \ldots, G_n\} \models^{L0} \mathcal{C}_i^{-1} \end{cases}$$

In that case, checking that contracts $\{\mathcal{C}_i\}_{i=1}^n$ L1-dominate a contract $\mathcal{C}$ requires one L1-satisfaction check and n L0-satisfaction checks. This is particularly interesting since checking L0-satisfaction may be achieved by using other tools or approaches (that may not need circular reasoning). Moreover, dominance can be established more often as L1-satisfaction is stronger than L0-satisfaction. Second:

Theorem 4 $K \sqsubseteq^{L0}_{A,gl} G \wedge E \sqsubseteq^{L1}_{G,gl} A$ *implies* $K \sqsubseteq^{L0}_{E,gl} G$.

This result made it possible in SPEEDS to incorporate results from tools checking L0-satisfaction with results obtained through L1-dominance (implemented by a set of L1-satisfaction checks), thus building a complete tool chain.

8.4 Conclusion and Future Work

The work presented in this chapter has been motivated by the necessity of combining contract-based verification tools and corresponding results for two component frameworks L0 and L1 defined in the context of the European SPEEDS project. In particular, we were interested in using dominance results established in L1 – and which cannot be obtained using the L0 refinement relation – for further reasoning in L0. To that purpose, we have presented an abstract notion of contract framework for a given component framework that defines three different notions of refinement, that is, conformance, dominance and satisfaction. We show how to derive these notions from refinement of closed systems and refinement under context and we provide a methodology for compositional and hierarchical verification of global properties.

We have studied circular reasoning as a powerful means for proving dominance. As circular reasoning does not always hold for usual notions of refinement, we provide proof rules for dominance relying on a relaxed notion of circular reasoning based on two notions of refinement. We have then shown that our abstract framework is general enough to represent both L0 and L1 as specific instances and proved that the L0 and L1 refinement relations satisfy the condition for relaxed circular reasoning.

This approach was applied to only simple case studies in the SPEEDS project and should therefore rather be seen as a proof of concept. The practical relevance of such an approach is that it opens up ways of connecting tools that work at different levels of abstraction, and relate their results to prove stronger properties. In addition, our results relax the requirements on the tools, since circular reasoning would not be needed at the L0 level.

Acknowledgments This work was supported in part by the EU projects COMBEST (n. 215543) and ArtistDesign (n. 214373).

References

1. Sangiovanni-Vincentelli, A., Damm, W., Passerone, R.: Taming Dr. Frankenstein: Contract-based design for cyber-physical systems. J. Control **18**(3), 217–238 (2012). doi:10.3166/EJC.18.217-238

2. Damm, W.: Controlling speculative design processes using rich component models. In: Proceedings of ACSD'05, pp. 118–119. IEEE Computer Society (2005)
3. Basu, A., Bozga, M., Sifakis, J.: Modeling heterogeneous real-time components in BIP. In: Proceedings of SEFM'06, pp. 3–12. IEEE Computer Society (2006)
4. Benveniste, A., Caillaud, B., Ferrari, A., Mangeruca, L., Passerone, R., Sofronis, C.: Multiple viewpoint contract-based specification and design. In: F.S. de Boer, M.M. Bonsangue, S. Graf, Willem-Paul de Roever (eds.) Formal Methods for Components and Objects, 6th International Symposium (FMCO 2007), Amsterdam, The Netherlands, October 24–26, 2007, Revised Papers, Lecture Notes in Computer Science, vol. 5382, pp. 200–225. Springer (2008). doi: 10.1007/978-3-540-92188-2
5. Benvenuti, L., Ferrari, A., Mangeruca, L., Mazzi, E., Passerone, R., Sofronis, C.: A contract-based formalism for the specification of heterogeneous systems. In: Proceedings of the Forum on Specification, Verification and Design Languages (FDL08), pp. 142–147. Stuttgart, Germany (2008). doi: 10.1109/FDL.2008.4641436
6. SPEEDS Consortium: Home page. http://www.speeds.eu.com
7. COMBEST Consortium: Home page. http://www.combest.eu
8. CESAR Consortium: Home page. http://www.cesarproject.eu/
9. Partners, S.: SPEEDS metamodel. SPEEDS project deliverable D2.1.5 (2009)
10. The Mathworks, Inc.: MATLAB simulink. http://www.mathworks.com
11. Alur, R., Henzinger, T.A.: Reactive modules. Formal Methods Syst. Des. **15**(1), 7–48 (1999)
12. Maier, P.: A lattice-theoretic framework for circular assume-guarantee reasoning. Ph.D. thesis, Universität des Saarlandes (2003)
13. Cobleigh, J.M., Avrunin, G.S., Clarke, L.A.: Breaking up is hard to do: An evaluation of automated assume-guarantee reasoning. ACM Trans. Softw. Eng. Methodol. **17**(2) (2008)
14. Alfaro, L., Henzinger, T.A.: Interface automata. In: Proceedings of ESEC/SIGSOFT FSE'01, pp. 109–120. ACM Press (2001)
15. Larsen, K.G., Nyman, U., Wasowski, A.: Interface input/output automata. In: Proceedings of FM'06, LNCS, vol. 4085, pp. 82–97 (2006)
16. Tripakis, S., Lickly, B., Henzinger, T.A., Lee, E.A.: On relational interfaces. In: Proceedings of EMSOFT'09, pp. 67–76 (2009)
17. Delahaye, B., Caillaud, B., Legay, A.: Probabilistic contracts: A compositional reasoning methodology for the design of stochastic systems. In: Proceedings of ACSD'10, pp. 223–232 (2010)
18. Raclet, J.B., Badouel, E., Benveniste, A., Caillaud, B., Legay, A., Passerone, R.: Modal interfaces: Unifying interface automata and modal specifications. In: Proceedings of the Ninth International Conference on Embedded Software (EMSOFT09), pp. 87–96. Grenoble, France (2009)
19. Raclet, J.B., Badouel, E., Benveniste, A., Caillaud, B., Passerone, R.: Why are modalities good for Interface Theories? In: Proceedings of the Ninth International Conference on Application of Concurrency to System Design (ACSD09), pp. 119–127. Augsburg, Germany (2009)
20. Raclet, J.B., Badouel, E., Benveniste, A., Caillaud, B., Legay, A., Passerone, R.: A modal interface theory for component-based design. Fundamenta Informaticae 108(1–2), 119–149 (2011). 10.3233/FI-2011-416
21. Larsen, K.G., Nyman, U., Wasowski, A.: Modal I/O automata for interface and product line theories. In: Proceedings of ESOP'07, LNCS, vol. 4421, pp. 64–79 (2007)
22. Quinton, S., Graf, S.: Contract-based verification of hierarchical systems of components. In: Proceedings of SEFM'08, pp. 377–381. IEEE Computer Society (2008)
23. Hafaiedh, I.B., Graf, S., Quinton, S.: Reasoning about safety and progress using contracts. In: Proceedings of ICFEM'10, pp. 436–451 (2010)
24. Graf, S., Passerone, R., Quinton, S.: Contract-based reasoning for component systems with complex interactions. Research report TR-2010-12, VERIMAG (2010 updated 2013)
25. Sifakis, J.: A framework for component-based construction. In: Proceedings of SEFM'05, pp. 293–300. IEEE Computer Society (2005)

26. Benveniste, A., Caillaud, B., Passerone, R.: A generic model of contracts for embedded systems. Rapport de recherche 6214, Institut National de Recherche en Informatique et en Automatique (2007)
27. Pinto, A., Bonivento, A., Sangiovanni-Vincentelli, A.L., Passerone, R., Sgroi, M.: System level design paradigms: Platform-based design and communication synthesis. ACM Trans. Des. Autom. Electron. Syst. **11**(3), 537–563 (2006). http://doi.acm.org/10.1145/1142980.1142982

Chapter 9
Extracting End-to-End Timing Models from Component-Based Distributed Embedded Systems

Saad Mubeen, Jukka Mäki-Turja and Mikael Sjödin

Abstract In order to facilitate the end-to-end timing analysis, we present a method to extract end-to-end timing models from component-based distributed embedded systems that are developed using the existing industrial component model, Rubus Component Model (RCM). RCM is used for the development of software for vehicular embedded systems by several international companies. We discuss and solve the issues involved during the model extraction such as extraction of timing information from all nodes and networks in the system and linking of trigger and data chains in distributed transactions. We also discuss the implementation of the method for the extraction of end-to-end timing models in the Rubus Analysis Framework.

9.1 Introduction

The model- and component-based development [1, 2] is often considered a promising choice for the development of distributed embedded systems for many reasons such as handling complexity of embedded software; lowering development costs; reducing time-to-market and time-to-test; allowing reusability; providing flexibility, maintainability and understandability; supporting modeling at higher level of abstraction and timing analysis during the process of system development. In distributed embedded systems with real-time requirements, the timing behavior of the system is as important as its functional behavior. The current trend for the industrial

S. Mubeen (✉) · J. Mäki-Turja · M. Sjödin
Mälardalen University, Västerås, Sweden
e-mail: saad.mubeen@mdh.se

M. Sjödin
e-mail: mikael.sjodin@mdh.se

J. Mäki-Turja
Arcticus Systems, Västerås, Sweden
e-mail: jukka.maki-turja@mdh.se

A. Sangiovanni-Vincentelli et al. (eds.), *Embedded Systems Development*, Embedded Systems 20, DOI: 10.1007/978-1-4614-3879-3_9,

development of such systems, especially in automotive domain, is focused towards handling timing related information and performing timing analysis as early as possible during the development process [3–5]. Hence, the component technology for the development of distributed embedded systems should support the extraction of required timing information into the end-to-end timing model.

Goals and Chapter Contribution. Our main goal is to extract the end-to-end timing models from component-based distributed embedded systems that are modeled with the existing industrial component model, i.e., the RCM [6, 7]. We focus on the following issues.

1. Extraction of timing information from all nodes and networks in a distributed embedded application into the end-to-end timing model.
2. Linking of trigger and data chains in distributed transactions, i.e., chains of tasks that are distributed over more than one node in a distributed embedded system.
3. Implementation of the timing model extraction method in the Rubus Analysis Framework.

Chapter Layout. The rest of the chapter is organized as follows. In Sect. 9.2, we discuss the background and research problem. In Sect. 9.3, we discuss main constituents of the end-to-end timing model. In Sect. 9.4, we discuss the model extraction method. Section 9.5 presents the related work. Section 9.6 concludes the chapter.

9.2 Background and Research Problem

9.2.1 The Rubus Concept

Rubus is a collection of methods and tools for model- and component-based development of dependable embedded real-time systems. Rubus is developed by Arcticus Systems [7] in close collaboration with several academic and industrial partners. Rubus is today mainly used for development of control functionality in vehicles by several international companies [8–11]. The Rubus concept is based around RCM and its development environment Rubus-Integrated Component development Environment (ICE) [7], which includes modeling tools, code generators, analysis tools and run-time infrastructure. The overall goal of Rubus is to be aggressively resource efficient and to provide means for developing predictable and analyzable control functions in resource-constrained embedded systems.

9.2.1.1 The Rubus Component Model

Rubus Component Model expresses the infrastructure for software functions, i.e., the interaction between the software functions in terms of data and control flow separately. The control flow is expressed by triggering objects such as clocks and events as well as other components. In RCM, the basic component is called Software Circuit (SWC). The execution semantics of an SWC are: upon triggering, read data on data *in-ports*; execute the function; write data on data *out-ports*; and activate the output trigger. RCM separates the control flow from the data flow among SWCs within a node. Thus, explicit synchronization and data access are visible at the modeling level. One important principle in RCM is to separate functional code and infrastructure implementing the execution model. RCM facilitates analysis and reuse of components in different contexts (SWC has no knowledge how it connects to other components). The component model has the possibility to encapsulate SWCs into software assemblies enabling the designer to construct the system at different hierarchical levels.

9.2.1.2 The Rubus Code Generator and Run-Time System

From the resulting architecture of connected SWCs, functions are mapped to run-time entities; tasks. Each external event trigger defines a task and SWCs connected through the chain of triggered SWCs (trigger chain) are allocated to the corresponding task. All clock triggered "chains" are allocated to an automatically generated static schedule that fulfills the precedence order and temporal requirements. Within trigger chains, inter-SWC communication is aggressively optimized to use the most efficient means of communication possible for each communication link. Allocation of SWCs to tasks and construction of schedule can be submitted to different optimization criterion to minimize, e.g., response times for different types of tasks, or memory usage. The run-time system executes all tasks on a shared stack, thus eliminating the need for static allocation of stack memory to each individual task.

9.2.1.3 The Rubus Analysis Framework

The Rubus model allows expressing real-time requirements and properties at the architectural level. For example, it is possible to declare real-time requirements from a generated event and an arbitrary output trigger along the trigger chain. For this purpose, the designer has to express real-time properties of SWCs, such as worst-case execution times and stack usage. The scheduler will take these real-time constraints into consideration when producing a schedule. For event-triggered tasks, response-time calculations are performed and compared to the requirements. The analysis supported by the model includes shared stack analysis [12] and distributed end-to-end response time and delay analysis [13].

9.2.1.4 The Rubus Simulation Model

The Rubus SIMulation Model (RSIM) and accompanying tools enable simulation and testing of applications modeled with RCM at various hierarchical levels such as an SWC or a function, a hierarchical RCM component structure as an Assembly (ASM), a complete Electronic Control Unit (ECU) application (may require I/O simulation), a set of ECU's, a distributed system (may require I/O simulation of each ECU). To verify the logical functionality of these objects, RSIM supports testing in an automatic generated framework based on the Rubus OS Simulator.

The input data is read from external tools or files, e.g., MATLAB, and fed to the simulation process that controls the stimulation of input ports and state variables using probes. The output from the simulation process is fed back to the external tools. By building a simulated environment around the application to be simulated, the execution of the application can be controlled from a high-level tool such as LabVIEW or MATLAB/Simulink. The high-level tools control the execution of the simulated target by means of commands to stop and run the target clock a specified number of ticks. The high-level tool sets the input data to the control function to be tested, performs a number of execution steps, and then reads the generated output data. In this way the execution flow can be visualized in each time increment.

9.2.2 Problem Statement: Linking of Distributed Chains

The distributed transactions in a distributed embedded system may consist of trigger chains, data chains or a combination of both. The first task (component) in a trigger chain is triggered independently, while the rest of the tasks are triggered by their respective predecessors as shown in Fig. 9.1a. Each task in a data chain is triggered independently as shown in Fig. 9.1b. A mixed chain is a combination of both trigger and data chains as shown in Fig. 9.1c. The end-to-end timing model should include linking and mapping information of all these chains. Moreover, the model should also identify the type of each chain because different timing constraints are specified on different types of chains [13].

The linking and mapping problem is common to all types of chains. For simplicity, we consider a distributed embedded system modeled with only trigger chains as shown in Fig. 9.2. There are two nodes in the system with three SWCs in node A and four SWCs in node B. SWCs communicate with each other by using both inter- and intra-node communication. The intra-node communication takes place via connectors whereas the inter-node communication takes place via a real-time network to which the nodes are connected. One trigger chain (distributed transaction) that is

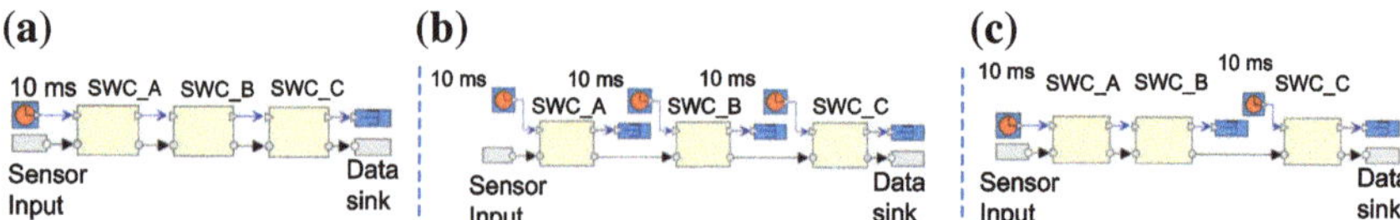

Fig. 9.1 Example of (**a**) Trigger chain (**b**) Data chain (**c**) Mixed chain

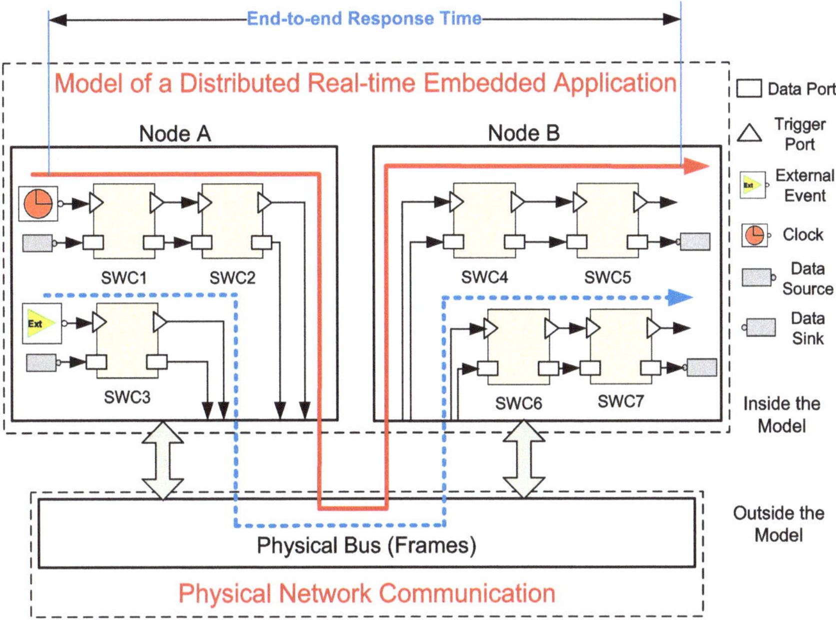

Fig. 9.2 Trigger chains in distributed transactions

activated by a clock consists of four Software Circuits, i.e., SWC1, SWC2, SWC4 and SWC5. It is identified with a solid-line arrow in Fig. 9.2. In this transaction, a clock triggers SWC1 which in turn triggers SWC2. SWC2 then sends a signal to the network. This signal is transmitted over the network in a message (frame) and is received by SWC4 at the receiver node. SWC4 processes it and sends it to SWC5. The elapsed time between the arrival of a triggering event at the input of the task corresponding to SWC1 and the production of response of the task corresponding to SWC5 is referred to as the holistic or end-to-end response time of the distributed transaction and is also identified in Fig. 9.2. The second trigger chain that is activated by an external event consists of three Software Circuits, i.e., SWC3, SWC6 and SWC7. It is identified by a broken-line arrow in Fig. 9.2.

There may not be direct triggering connections between any two neighboring SWCs in the chain which is distributed over more than one node, e.g., SWC2 and SWC4 in Fig. 9.2. In this case, SWC2 communicates with SWC4 by sending signals via the network. Here, the problem is that when a trigger signal is produced by SWC2, it may not be sent straightaway as a message on the network. A message may combine several signals and hence, there may be some waiting time for the signal to be sent on the network. The message may be sent periodically or sporadically or by any other rule defined by the underlying network protocol. When such trigger chains are modeled using the component-based approach, it is not straightforward to link them to extract the end-to-end timing model. For example, if a message is received at node B then the following information should be available to correctly link the received message in the chain: the ID of the sender node; the ID of the task that generated this message; the ID of the destination node; and the ID(s) of the task(s) that should receive this message. In order to get a bounded end-to-end delay, a more important question is when and who will trigger the destination SWC when a message is received at node B.

The existing modeling components in RCM do not provide enough support to link and extract the corresponding timing information of distributed chains. Therefore, special objects in the component technology are needed to provide the linking information of distributed chains to extract end-to-end timing information. Further, there is a need to model mapping between signals and messages and vice versa. SWCs inside a node communicate via signals whereas they communicate via messages if located on different nodes in a distributed transaction. Moreover, there is a need to model exit and entry points for RCM models. An exit point is where a message (data) leaves the model and is transmitted according to the protocol-specific rules of the network. Similarly, an entry point is where a message enters the model from the model of the network or any other model. The reason for the need of modeling exit and entry points for RCM models is to get the bounded delays in distributed transactions. The model of entry and exit points will support the use of nodes developed using RCM with the nodes developed by other component technologies.

9.3 End-to-End Timing Model

The end-to-end timing model consists of timing properties, requirements and dependencies concerning all tasks, messages, task chains and distributed transactions in a distributed embedded system under analysis. Basically, it consists of two models, i.e., system timing model and system linking model. All the required timing information of each node in a distributed embedded application is extracted into a node timing model. Similarly, the timing information of all networks in a distributed embedded application is extracted into a network timing model. Together the node and network timing model comprise the system timing model. All mapping and linking information of distributed chains is extracted into the system linking model.

9.3.1 System Timing Model

Node Timing Model. The node timing model contains node-level timing information. It is based on a transaction model with offsets developed by [14] and later on, extended by many researchers, e.g., [15, 16]. A node, Γ, consists of a set of k transactions $\Gamma_1, \ldots, \Gamma_k$. Each transaction Γ_i is activated by mutually independent events, i.e., the phasing between them is arbitrary. The activating events can be a periodic sequence of events with a period T_i. In case of sporadic events, T_i denotes the minimum inter-arrival time between two consecutive events.

There are $|\Gamma_i|$ tasks in a transaction Γ_i and each task may not be activated until a certain time, called an *offset*, elapses after the arrival of the external event. By task activation we mean that the task is released for execution. A task is denoted by τ_{ij}. The first subscript, *i*, specifies the transaction to which this task belongs and the second subscript, *j*, denotes the index of the task within the transaction.

A task, τ_{ij}, is defined by the following attributes: a priority (P_{ij}), a worst-case execution time (C_{ij}), an offset (O_{ij}), maximum release jitter (J_{ij}), an optional deadline (D_{ij}), maximum blocking time which is the maximum time the task has to wait for a resource that is locked by a lower priority task (B_{ij}). In order to calculate the blocking time for a task, usually, a resource locking protocol like priority ceiling or immediate inheritance is used. Each task has a worst-case response time denoted by R_{ij}. In this model, there are no restrictions placed on offset, deadline or jitter, i.e., they can each be either smaller, greater or equal to the period.

Network Timing Model. This model contains network-level timing information of a distributed embedded system. A network consists of a number of nodes that are connected through a real-time network. Currently, the model supports Controller Area Network (CAN) and its higher-level protocols such as CANopen, CAN for Military Land Systems domain (MilCAN) and Hägglunds CAN (HCAN) [17]. If a task on one node intends to communicate with a task on another node, it queues a message in the send queue of its node. The network communication protocol ensures the arbitration and transmission of all messages over the network.

Each message *m* has the following attributes: a unique identifier (ID_m); transmission type showing whether the message is periodic or sporadic or mixed [17]; a unique priority (P_m); transmission time (C_m); release jitter (J_m) which is inherited from the difference between the worst- and best-case response times of the task queueing the message; data payload (s_m) in the message; period (T_m) in the case of periodic transmission, Minimum Update Time (MUT_m) which is the minimum time that should elapse between the transmission of any two sporadic messages in the case of sporadic transmission, or both T_m and MUT_m in the case of mixed transmission [17–20]; blocking time (B_m) which is the maximum time a message can be blocked by lower priority messages; and worst-case response time (R_m).

9.3.2 System Linking Model

In distributed embedded systems, the transactions are usually distributed over several nodes. Hence, there exist chains of components (tasks) that may be distributed over more than one node. A task chain consists of a number of tasks that are in a sequence and have one common ancestor. A task in a chain may receive trigger, data or both from its predecessor. Two neighboring tasks in a distributed transaction may reside on two different nodes, while the nodes communicate with each other via a network. When there are chains in a distributed embedded system, the end-to-end timing model should not only contain timing related information but also the linking information among all tasks and messages within each distributed chain (see Sect. 9.2.2) . The extraction of linking information from chains in a distributed real-time system is more complex compared to a single node real-time system.

9.4 Extraction of End-to-End Timing Model

In this section, we resolve the issues discussed in the previous section. We also show the applicability of our approach by modeling a two-node distributed embedded application with RCM. Finally, we present the conceptual organization of the method for the extraction of end-to-end timing models in Rubus-ICE.

9.4.1 Proposed Solution

9.4.1.1 Addition of Special Components in RCM

In order to model real-time network communication and legacy communication in distributed embedded systems, we introduced special purpose Software Circuits in RCM, i.e., Output Software Circuit (OSWC) and Input Software Circuit (ISWC) in [21]. There is one OSWC for each message that a node sends to the network. Similarly, there is one ISWC for each message that a node receives from the network. It should be noted that each of these special-purpose components is translated to an individual task at run time. We also introduced a new object in RCM, i.e., the Network Specification (NS) that represents the model of communication in a physical network [21]. There is one NS for each network protocol. NS contains Signal Mapping which includes the following information: How are signals mapped to messages? How many signals a message contains? How are signals encoded in a message at the sender node? How are signals decoded from a message at the receiving node? The model representation of OSWC, ISWC and NS in a two-node distributed embedded system modeled with RCM is shown in Fig. 9.3.

When the OSWC component is triggered, it executes the required functionality (e.g., mapping of signals to a message) and then the data (message) is transferred from the RCM model of a node to the network controller or another model of communication network. Therefore, OSWC also represents the model of an exit point for RCM models. Similarly, ISWC component also represents the model of an entry point for RCM models. Since the trigger *in-ports* of all OSWC components and trigger *out-ports* of all ISWC components along a distributed transaction are referenced in NS, the end-to-end timing delay can be bounded by specifying the delay in the extra-model medium.

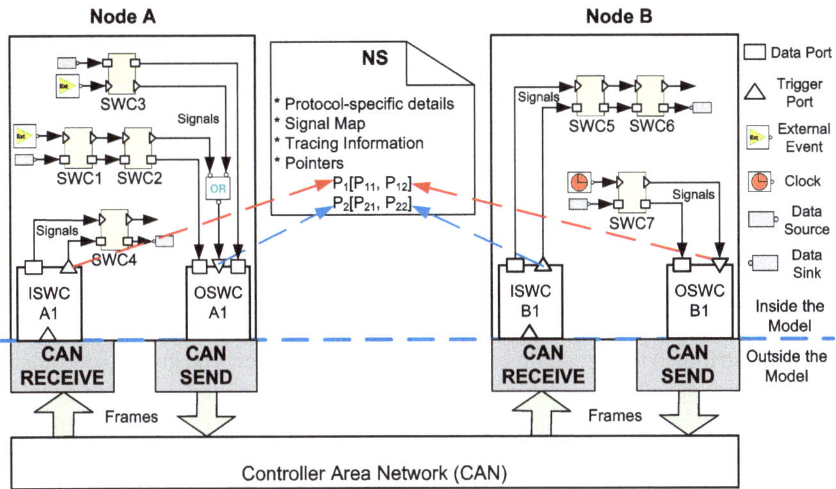

Fig. 9.3 Example of a two-node distributed embedded system modeled with RCM

9.4.1.2 Identification of Distributed Chains

In order to unambiguously identify each individual chain, we attach "trigger dependency" attribute with each task. This attribute is part of the data structure of tasks in the end-to-end timing model. If a task is triggered by an independent source such as a clock then this attribute will be assigned "independent". On the other hand, if the task is triggered by another task then this parameter will be assigned "dependent". Moreover, a precedence constraint will also be specified on this task in the case of dependent triggering. If this attribute for all tasks (except the first) has value "dependent", the chain will be identified as a trigger chain. On the other hand, if this attribute for more than one task in a chain has "independent" value then the chain will identified as a data chain.

9.4.1.3 Linking and Mapping of Distributed Chains

The linking information of all distributed chains in the modeled distributed embedded application is provided in the Network Specification. We assign pointers (references) to trigger *in-ports* of OSWCs and the trigger *out-ports* of ISWCs along the same distributed transaction. All such pointers for all trigger chains in the system are specified in the NS.

An example of a two-node distributed embedded system modeled in RCM is shown in Fig. 9.3. There are four SWCs in Node A while three SWCs in Node B. We consider CAN or any of its high-level protocols for inter-node communication. In this example, the nodes are connected to a CAN network. There are three trigger chains in the system that are distributed over two nodes:

- $EC_1 : SWC1 \rightarrow SWC2 \rightarrow OSWC\_A1 \rightarrow ISWC\_B1 \rightarrow SWC5 \rightarrow SWC6$.
- $EC_2 : SWC3 \rightarrow OSWC\_A1 \rightarrow ISWC\_B1 \rightarrow SWC5 \rightarrow SWC6$.
- $EC_3 : SWC7 \rightarrow OSWC\_B1 \rightarrow ISWC\_A1 \rightarrow SWC4$.

The trigger chains EC_1 and EC_2 are triggered by external events whereas the trigger chain EC_3 is triggered by a clock. The references to the trigger ports of OSWC and ISWC in each trigger chain are specified in NS. There is a pointer array P_1 that references the trigger *in-port* of OSWC B1 in Node B and trigger *out-port* of ISWC A1 in node A. Similarly, a pointer array P_2 is stored in NS that points to the trigger *in-port* of OSWC A1 in Node A and trigger *out-port* of ISWC B1 in node B. In this way, all the neighboring components located in different nodes within a distributed trigger chain can be linked to each other. The grey boxes outside the model are specific for each network communication protocol. In this example they represent CAN SEND and CAN RECEIVE routines. The CAN SEND grey box represents a CAN controller in a node and is responsible for receiving messages from the corresponding OSWC and queueing them for transmission over the network.

When a message arrives at the receiving node, it is transferred by the physical network drivers to the CAN RECEIVE grey box which is responsible for raising an interrupt request and passing the message to the corresponding ISWC component. In this case, the *TrigInterrupt* object in RCM corresponding to the interrupt is connected to the *in-port* of the ISWC component. If CAN drivers use polling-based processing instead of interrupts then the *in-port* of the ISWC component is connected to the clock object in RCM whose period is equal to the polling period. Upon receiving a message, an ISWC component decodes it, extracts signals from it, places the data on the corresponding data port (connected to the data *in-port* of the destination SWC) and triggers the corresponding trigger port by using the linking information in NS. It should be noted that there can be more than one ISWC and OSWC components in a node. It can be seen from Fig. 9.3 that OSWC and ISWC essentially make the exit and entry points for the node models.

9.4.2 Extraction of End-to-End Timing Model in Rubus-ICE

In Rubus-ICE, a distributed embedded application is modeled in Rubus Designer. It is then compiled to the Intermediate Compiled Component Model (ICCM). Apart from the compiled component model, ICCM file also includes timing and linking information of the modeled system. The end-to-end timing model that is implemented in the Rubus Analysis Framework, extracts the required timing and linking information from ICCM file as shown in Fig. 9.4. The end-to-end timing model consists of three models, i.e., node timing model, network timing model and system linking model. From the extracted timing model, the Rubus Analysis Framework performs the end-to-end timing analysis and then provides the results, i.e., response times of individual tasks, response times of network messages, end-to-end response times and delays of distributed chains, network utilization, etc., back to the Rubus-ICE tool suite. The

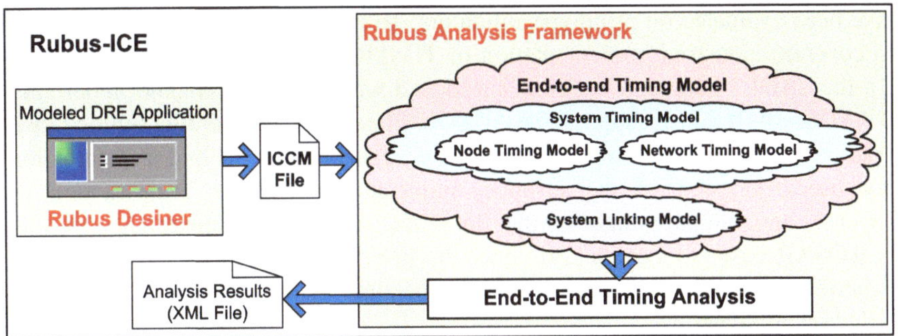

Fig. 9.4 Extraction of end-to-end timing model in Rubus tool-suite

schedulability analysis results of a case study, performed using our timing model extraction method, are presented in [13].

9.5 Related Work

There are very few commercial component models for the development of distributed embedded systems especially in automotive domain. In our previous work, we carried out a detailed comparison of RCM with various models for distributed embedded systems [21]. We briefly highlight a few of them. AUTomotive Open System ARchitecture (AUTOSAR) [22] is a standardized software architecture for the development of software in automotive domain. It can be viewed as a standardized distributed component model [23]. When AUTOSAR was being developed, there was no focus placed on its ability to specify and handle real-time requirements and properties. On the other hand, such requirements and capabilities were strictly taken into account right from the beginning during the development of RCM. AUTOSAR describes embedded software development at a relatively higher level of abstraction compared to RCM. A Software Circuit in RCM more resembles to a runnable entity (a schedulable part of AUTOSAR software component) instead of AUTOSAR software component. As compared to AUTOSAR, RCM clearly distinguishes between control flow and data flow among software components in a node. AUTOSAR hides the modeling of execution environment. Whereas, RCM explicitly allows the modeling of execution requirements, e.g., jitter and deadlines, at an abstraction level close to the functional modeling while abstracting the implementation details.

TIMing MOdel (TIMMO) [5] project is an initiative to provide AUTOSAR with a timing model. The timing extensions proposed in this project are included in the version 4.0 of AUTOSAR specification [24]. It describes a predictable methodology and a language, Timing Augmented Description Language (TADL) [4], to express timing requirements and timing constraints in all design phases during the development of automotive embedded systems. Both TIMMO methodology and TADL have been evaluated on prototype validators. To the best of our knowledge there is no concrete industrial implementation of TIMMO. In the TIMMO-2-USE project [3], the TADL2 language has been introduced which includes a major redefinition of TADL. TADL2 also supports the AUTOSAR extensions regarding timing model. Apart from the redefinition of the language, algorithms, tools and a methodology have been developed to model advanced timing at different levels of abstraction. The use cases and validators indicate that the project results are in compliance with the AUTOSAR-based tool chain [24]. Since this project is recently finished, it may take some time for the results of the project to become mature and find their way into industrial applications.

ProCom [25] is a two-layer component model for the development of distributed embedded systems. ProCom is inspired by RCM, and there are a number of similarities between the ProSave modeling layer (a lower layer in ProCom) and RCM. For example, components in both ProSave and RCM are passive. Similarly, both models clearly separate data flow from control flow among their components. Moreover, the communication mechanism for component interconnection used in both models is pipe-and-filter. The validation of a complete distributed embedded system, modeled with ProCom, is yet to be done. Moreover, the development environment and the tools accompanying ProCom are still evolving.

BIP framework [26] provides a 3-layered representation, i.e., behavior, interaction and priority for modeling of heterogeneous real-time components. Unlike RCM, it does not distinguish between required and provided or input and output interfaces (or ports). BIP uses triggering, rendezvous and broadcast styles for component interaction, while RCM uses the pipe-and-filter style for interaction among components. BIP provides connections to IF Toolset [27] and PROMETHEUS tools [28] to support modeling, validation, static analysis, model checking and simulation. On the other hand, RCM is supported by the Rubus-ICE tool suite that provides a complete modeling, analysis and simulation support for the component-based development of distributed real-time systems.

Robocop [29, 30] is a component-based framework for the development of middleware in the consumer electronics domain, e.g., consumer devices like mobile phones, DVD players, ATM machines, etc. On the other hand, RCM is intended for the development of real-time embedded systems in the automotive domain. The interaction mechanism among components in both models is also different as Robocop uses request response while Rubus uses pipe and filter. However, there are several similarities between Robocop and RCM, e.g., both have a low resource footprint, both are able to generate C code and both support deployment at the compilation step (in addition, Robocop supports deployment at run time).

A related research presents a detailed overview of timing aspects during the design activities of automotive embedded systems [31]. In [32], the authors define end-to-end delay semantics and present a formal framework for the calculation of end-to-end delays for register-based multi-rate systems. Like any other timing analysis, they assume that the timing information of the system is available as an input. On the other hand, our focus is on the extraction of required information into an end-to-end timing model to carry out end-to-end timing analysis.

In our previous work, we extended RCM to support modeling and analysis of distributed embedded systems. In [33], we explored various options for modeling of real-time network communication in RCM. In [21], we discussed modeling of legacy communication in component-based distributed embedded systems. We added new components in RCM to encapsulate and abstract the communication protocols, allow use of legacy nodes and legacy protocols in a component- and model-based software engineering environment, and support model- and component-based development of new nodes that are deployed in legacy systems that use predefined communication rules. Further, we highlighted the problem of linking trigger chains in the transactions that are distributed over several nodes in a distributed embedded system [34].

9.6 Conclusion

In this chapter, we discuss the extraction of end-to-end timing models from component-based distributed embedded systems modeled with the industrially available component model, the Rubus Component Model (RCM). The purpose of extracting such models is to perform the end-to-end timing analysis during the development process. We discussed and resolved various issues during the model extraction such as extraction of timing information from all nodes and networks in the system and extraction of linking model containing the linking information of all distributed chains. We also described the implementation of the end-to-end timing model extraction method in the Rubus Analysis Framework.

Although, we discussed the extraction of end-to-end timing models from RCM models, we believe that the model extraction method is also suitable for other component models for the development of distributed embedded systems that use a pipe-and-filter style for component interconnection, e.g., ProCom [25], COMDES [35]. Moreover, our approach can be used for any type of "inter-model signaling", where a signal leaves one model (e.g. a node, or a core, or a process) and appears again in some other model.

Acknowledgments This work is supported by the Swedish Knowledge Foundation (KKS) within the project FEMMVA. The authors thank the industrial partners Arcticus Systems, BAE Systems Hägglunds and Volvo Construction Equipment (VCE), Sweden.

References

1. Crnkovic, I., Larsson, M.: Building Reliable Component-Based Software Systems. Artech House, Inc., USA (2002)
2. Henzinger, T.A., Sifakis, J.: The embedded systems design challenge. In: Proceedings of the 14th international symposium on formal methods (FM), Lecture Notes in Computer Science, pp. 1–15. Springer, Heidelberg (2006)
3. TIMMO Consortium: TIMMO-2-USE. http://www.timmo-2-use.org/

4. TIMMO Consortium: TADL: Timing Augmented Description Language, Version 2. TIMMO (TIMing MOdel), Deliverable 6 (2009)
5. TIMMO Consortium: TIMMO Methodology, Version 2. TIMMO (TIMing MOdel), Deliverable 7. The TIMMO Consortium (2009)
6. Hänninen K., et al.: The rubus component model for resource constrained real-time systems. In: 3rd IEEE International Symposium on Industrial Embedded Systems (2008)
7. Arcticus Systems AB: Arcticus Systems home page. http://www.arcticus-systems.com
8. BAE Systems: BAE Systems Hägglunds. http://www.baesystems.com/hagglunds
9. Volvo AB: Volvo Construction Equipment. http://www.volvoce.com
10. Mecel AB: Home page. http://www.mecel.se
11. Knorr-Bremse AG: Home page. http://www.knorr-bremse.com
12. Hänninen, K.: Efficient memory utilization in resource constrained real-time systems. Ph.D. thesis, Mälardalen University, Sweden (2008)
13. Mubeen, S., Mäki-Turja, J., Sjödin, M.: Support for end-to-end response-time and delay analysis in the industrial tool suite: Issues, experiences and a case study. Comput. Sci. Inf. Sys. **10**(1), 453–482, ISSN: 1361–1384, (2013)
14. Tindell, K.: Adding time-offsets to schedulability analysis. Department of Computer Science, University of York, England, Tech. rep (1994)
15. Palencia, J., Harbour, M.G.: Schedulability analysis for tasks with static and dynamic offsets. Real-Time Systems Symposium, IEEE International p. 26 (1998). URLhttp://doi.ieeecomputersociety.org/10.1109/REAL.1998.739728
16. Mäki-Turja, J., Nolin, M.: Efficient implementation of tight response-times for tasks with offsets. Real-Time Syst. **40**(1), 77–116 (2008). URLhttp://dx.doi.org/10.1007/s11241-008-9050-9
17. Mubeen, S., Mäki-Turja, J., Sjödin, M.: Extending schedulability analysis of controller area network (CAN) for mixed (periodic/sporadic) messages. In: 16th IEEE Conference on Emerging Technologies and Factory Automation (ETFA) (2011). doi:10.1109/ETFA.2011.6059010.
18. Mubeen, S., Mäki-Turja, J., Sjödin, M.: Extending response-time analysis of controller area network (CAN) with FIFO queues for mixed messages. In: 16th IEEE Conference on Emerging Technologies and Factory Automation (ETFA), pp. 1–4 (2011). doi:10.1109/ETFA.2011.6059188.
19. Mubeen, S., Mäki-Turja, J., Sjödin, M.: Response-time analysis of mixed messages in controller area network with priority- and FIFO-queued nodes. In: 9th IEEE International Workshop on Factory Communication Systems (WFCS) (2012)
20. Mubeen, S., Mäki-Turja, J., Sjödin, M.: Worst-case response-time analysis for mixed messages with offsets in controller area network. In: 17th IEEE Conference on Emerging Technologies and Factory Automation (ETFA) (2012)
21. Mubeen, S., Mäki-Turja, J., Sjödin, M., Carlson, J.: Analyzable modeling of legacy communication in component-based distributed embedded systems. In: 37th Euromicro Conference on Software Engineering and Advanced Applications (SEAA), pp. 229–238 (2011). doi:10.1109/SEAA.2011.43
22. AUTOSAR Consortium: AUTOSAR Technical Overview, Version 2.2.2. AUTOSAR - AUTomotive Open System ARchitecture, Release 3.1, The AUTOSAR Consortium, Aug (2008). http://autosar.org
23. Heinecke, H., et al.: AUTOSAR - Current results and preparations for exploitation. In: Proceedings of the 7th Euroforum Conference, EUROFORUM '06 (2006)
24. TIMMO Consortium: Mastering Timing Information for Advanced Automotive Systems Engineering - In the TIMMO-2-USE Brochure (2012). http://www.timmo-2-use.org/pdf/T2UBrochure.pdf
25. Sentilles, S., Vulgarakis, A., Bures, T., Carlson, J., Crnkovic, I.: A Component Model for Control-Intensive Distributed Embedded Systems. In: 11th International Symposium on Component Based Software Engineering (CBSE2008), pp. 310–317. Springer, Heidelberg (2008)
26. Basu, A., Bozga, M., Sifakis, J.: Modeling heterogeneous real-time components in BIP. In: Proc. of SEFM'06, pp. 3–12. IEEE Computer Society (2006)

27. Bozga, M., et. al.: The IF Toolset. In: Formal Methods for the Design of Real-Time Systems, Lecture Notes in Computer Science, vol 3185, pp. 237–267. Springer, Hedielberg (2004)
28. Gssler, G.: Prometheus - A Compositional Modeling Tool for Real-Time Systems.In, Workshop on Real-Time Tools (RT-TOOLS) (2001). (2001)
29. ROBOCOP Team: ROBOCOP project. http://www.hitech-projects.com/euprojects/robocop/deliverables.htm
30. Muskens, J., Chaudron, M.R.V., Lukkien, J.J.: A component framework for consumer electronics middleware. In: Component-Based Software Development for Embedded Systems. pp 164–184, (2005)
31. Scheickl, O., Rudorfer, M.: Automotive real time development using a timing-augmented AUTOSAR specification. ERTS, (2008)
32. Feiertag, N., Richter, K., Nordlander, J., Jonsson, J.: A compositional framework for end-to-end path delay calculation of automotive systems under different path semantics.In:Workshop on Compositional Theory and Technology for Real-Time Embedded Systems (CRTS)(2008)
33. Mubeen, S., Mäki-Turja, J., Sjödin, M.: Exploring Options for Modeling of Real-Time Network Communication in an Industrial Component Model for Distributed Embedded Systems. In: The 6th International Conference on Embedded and Multimedia Computing (EMC-2011), Lecture Notes in Electrical Engineering, vol. 102, pp. 441–458. Springer Berlin / Heidelberg (2011)
34. Mubeen, S., Mäki-Turja, J., Sjödin, M.: Tracing event chains for holistic response-time analysis of component-based distributed real-time systems. In: 23rd Euromicro Conference on Real-Time Systems (ECRTS 2011), WIP Session. ACM SIGBED, Review (2011)
35. Ke, X., Sierszecki, K., Angelov, C.: COMDES-II: A Component-Based Framework for Generative Development of Distributed Real-Time Control Systems. In: 13th IEEE International Conference on Embedded and Real-Time Computing Systems and Applications (RTCSA), 2007, pp. 199–208 (2007). doi:10.1109/RTCSA.2007.29

Part IV
Timing Analysis and Time-Based Synthesis

Chapter 10
Distributed Priority Assignment in Real-Time Systems

Moritz Neukirchner, Steffen Stein and Rolf Ernst

Abstract Recent advances in in-system performance analysis allow to determine feasibility of a system configuration within the system itself. Such methods have been successfully used to perform admission control for updates in distributed real-time systems. Parameter synthesis, which is necessary to complement the admission control with self-configuration capabilities, lags behind because current approaches cannot be distributed properly or due to necessary design-time preprocessing steps. In this chapter we present a distributed algorithm to find feasible execution priorities in distributed static-priority-preemptively (SPP) scheduled real-time systems under consideration of end-to-end path latencies. The presented algorithm builds on top of an existing distributed feasibility test, which is derived from compositional performance analysis [1]. With an extensive set of pseudo-randomly generated testcases we demonstrate the applicability of the approach and show that the proposed algorithm can even compete with state-of-the-art design time tools at a fraction of the run time. Thus, despite its application to admission control, the approach is generally applicable to the problem of scheduling priority assignment.

10.1 Introduction

The integration of several components to a complex real-time system, such as an automotive platform, is a challenging task, as the integration process may introduce non-functional dependencies among otherwise independent components. Such dependencies arise through e.g. use of a common communication bus. Correct functioning is usually assured through means of extensive testing and formal verification. When such systems are being updated or extended after deployment, the feasibility

M. Neukirchner (✉) · S. Stein · R. Ernst
Institute of Computer and Network Engineering, Technische Universität Braunschweig, Braunschweig, Germany
e-mail: neukirchner@ida.ing.tu-bs.de

A. Sangiovanni-Vincentelli et al. (eds.), *Embedded Systems Development*, Embedded Systems 20, DOI: 10.1007/978-1-4614-3879-3_10,

of the system has to be re-verified considering the changes imposed through the update.One option to address this problem is by admission control mechanisms, that perform a formal verification in the system itself [2]. In this case the feasibility of the update is analyzed prior to system reconfiguration/update and infeasible changes are rejected.

However, in real-time systems feasibility of a system configuration heavily depends on the assignment of scheduling parameters. Thus, when updating a system, by e.g. adding a new software component, the update may be deemed infeasible by the admission control, although it may be feasible under a different assignment of scheduling parameters. To address this issue, we propose to extend admission control by a self-configuration service, which reassigns scheduling parameters such that configurations, that would otherwise be rejected, can be allowed to execute.

Specifically, we address the constraint satisfaction problem of finding feasible priority assignments in SPP scheduled real-time systems under consideration of end-to-end path latency constraints. The algorithm that we present relies on a distributed implementation of compositional performance analysis [1, 3], which has been successfully used for a distributed admission control scheme [2]. In order to reduce runtime overhead and to integrate with the admission control scheme the proposed algorithm is also implemented distributedly. Despite its application to the above admission control scheme, the approach is generally applicable to the problem of distributed priority assignment and not bound to any specific framework. As we show later, the algorithm can even outperform a state of the art design time tool.

The remainder of this work is structured as follows. First we will review previous approaches for priority assignment in real-time systems (Sect. 10.2). We will then provide a brief description of the system model and provide some insight on the underlying admission control scheme and its distributed feasibility test [1] (Sect. 10.3). In Sect. 10.4 we will outline the general strategy for the distributed self-configuration algorithm and Sect. 10.5 introduces the *local improvement target*, which forms the basic metric used for priority assignment. Section 10.6 then describes the proposed algorithm for distributed priority assignment, which integrates into the general strategy. We evaluate its performance with two benchmark algorithms (Sect. 10.7). Section 10.8 concludes the chapter.

10.2 Related Work

The problem of priority assignment has been studied intensively in the scope of scheduling analysis. First approaches addressed uni-processor systems and the question of schedulability of periodic tasks with task deadlines equal to their periods [4]. Later work reduced the restrictions on task deadlines [5, 6] and on periodicity [7, 8]. Extensions to multi-processor systems were then presented in [9–11]. In our scenario of admission control and self-configuration we consider tasks with communication dependencies and constraints on end-to-end path latencies. As the above approaches consider independent tasks and task- rather than path-latencies they are of limited applicability.

Hamann et al. [12] and Glaß et al. [13] both presented frameworks for design-space exploration of real-time systems that do not pose these restrictions on the system model. Both approaches use a genetic algorithm (GA) and a tool for performance analysis [3] to explore the design space. They support a multitude of parameters for optimization, among which is priority assignment. In this work we will use [12] as a benchmark. Genetic algorithms are generally computationally expensive due to the large number of individuals that are required to derive a solution. This is undesirable for use in self-configuration in resource-constrained embedded systems.

Another approach that is specifically targeted at runtime assignment of scheduling parameters was presented by [14, 15]. Here, a control-theoretic approach is taken to dynamically adjust scheduling parameters based on the actual workload of the system. This approach, however, is only suitable for soft real-time systems and cannot be applied if hard constraints have to be considered.

An approach that is more suitable for use in in-system admission control, that shall ensure adherence to hard constraints, is to divide end-to-end deadlines into local deadlines. Based on local algorithms tasks are then scheduled w.r.t. their local deadlines. Jonsson and Shin [16] provides a good overview of work following this approach. While most of these approaches target design-time optimization, e.g. [17], the algorithm presented in [18] aims to find feasible schedules in-system. However, the calculation of the local deadlines has to be performed in an offline pre-processing step, which can significantly limit the exploitation of available system slack.

In [19] a distributed algorithm for local deadline assignment in earliest deadline first (EDF) scheduled systems is presented. This approach can also be used for priority assignment in systems with static priorities. However, it is not applicable to systems which contain cyclic scheduling dependencies.

Neukirchner et al. [20] presented a distributed heuristic priority assignment algorithm, that does not require division of path latency deadlines into local task deadlines, while still allowing an efficient distributed implementation. The algorithm presented in this chapter builds on the same distribution approach as [20] (see Sect. 10.4). However, as will be shown in Sect. 10.7, our algorithm provides greatly improved results and even outperforms the design-time solution [12], which was chosen as benchmark.

10.3 System Model and Admission Control Concept

In this section, we introduce the system model and admission control concept, which forms the basis for our algorithm.

We use the system model as in [3]. In this system model a hardware *platform* $\mathcal{P}$ consists of multiple processors interconnected by communication media. We will refer to processors and communication media as (computational and communication) *resources* ρ_j. On this platform a set of communicating *tasks* $\Gamma = \{\tau_i\}$ are executed. A set of paths $\Psi = \{\psi_k\}$ with constraints on end-to-end latencies $C = \{\chi_{\psi_k} : \psi_k \in \Psi\}$ are specified for the task set.

The distributed priority assignment algorithm presented in this chapter relies on compositional performance analysis (CPA) [3, 21] as feasibility test, while also other modular analyses as e.g. Modular Performance Analysis (MPA) [22] can be used. CPA composes local schedulability analysis algorithms using event model interfaces. Schedulability analysis algorithms derive worst-case response times from worst-case execution times of tasks for a given scheduling policy. Algorithms exist for a multitude of scheduling and bus arbitration schemes, e.g. for static priority preemptive scheduling [23], Round Robin [24], or CAN Bus [25]. Using these functions, CPA derives bounds on the individual response time of each task in the system also under the assumption of communicating tasks. The response times are aggregated to compute bounds on path latencies [26].

Stein et al. [1] has presented a distributed performance analysis (DPA) based on CPA, which can be used for admission control within a real-time system. The distributed priority assignment algorithm, which we present in this chapter, is designed to be used as self-configuration service along with this DPA-based admission control scheme. The DPA implementation is composed of several DPA instances, one residing at each resource in the system. The single instances communicate the worst-case timing behavior of their tasks and cooperatively determine worst-case system level timing. Each DPA instance only contains model data of tasks that reside on the resource of that DPA. We refer to this information as the *local model* of the DPA instance. Specifically, a DPA instance can provide information on worst-case task response times ω_{τ_i} (WCRT), worst-case path latencies λ_{ψ_k} and path latency constraints $\chi_{\psi_k} \in C$ of tasks that reside on the same resource as the DPA instance. The provided estimations on WCRTs are *monotonic*, i.e. if a task's priority is increased/decreased and all other parameters remain equal, its worst-case response time can only decrease/increase or remain equal, respectively.

To be able to reason about data within the local model of a DPA instance we introduce some sets of variables. Let Γ_{ρ_j} be the set of tasks that are mapped on resource ρ_j and Γ_{ψ_k} be the set of tasks that are part of path ψ_k. Furthermore let Ψ_{ρ_j} be the set of paths that have at least one task mapped on resource ρ_j ($\tau_i \in \Gamma_{\rho_j}$) and Ψ_{τ_i} be the set of paths that task τ_i is part of ($\tau_i \in \Gamma_{\psi_k}$). These definitions are later required to define the self-configuration algorithm.

10.4 Self-Configuration Strategy

In this section we present the general approach to self-configuration.The distributed self-configuration (DSC) algorithm relies on the model-based Distributed Performance Analysis (DPA) algorithm [1] which is used for admission control. Each DPA instance is complemented by a DSC instance - both instances residing at the same resource (Fig. 10.1a). The DSC can request estimations on WCRTs and path latencies from the DPA. Due to the distribution of the model and the DPA, each DSC

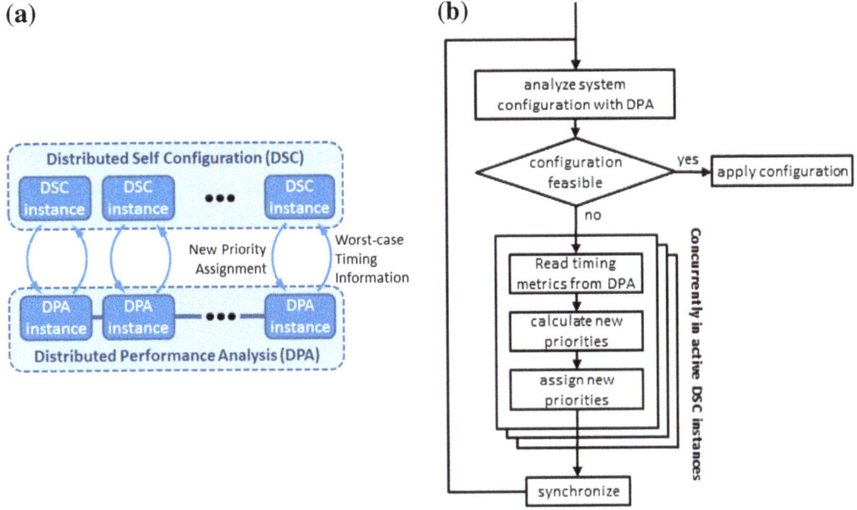

Fig. 10.1 General algorithm flow **a** architecture view. **b** logic view

instance can only access the information provided by its attached DPA instance, i.e. the DPA's local model. Each DSC instance can reassign task priorities in the local model of its attached DPA instance. While the DPA instances communicate to analyze a system configuration, the DSC instances do not require to communicate except for synchronization.

Fig. 10.1b shows the DSC flow. The DPA analyzes the system model. A DSC instance becomes active when its DPA instance reports a constraint violation on its respective resource, i.e. if the worst-case path latency of any path on that resource exceeds its constraint. Based on local rules and data available from their attached DPA instances all active DSC instances concurrently compute new priority assignments and insert them into the model of the DPA. All active DSC instances synchronize (e.g. using a barrier synchronization protocol as described in [27]) to ensure a consistent model. Then the DPA analyzes the modified configuration again. This loop is executed on a resource whenever the current priority assignment does not satisfy all path latency constraints. Each execution of this loop is referred to as a *DSC step*. As the DPA is performed synchronized across all affected resources, DPA and DSC are performed in a lock-step manner. If a global solution is found (i.e. all constraints are satisfied), a feasible configuration has been found and the update to the system configuration can be accepted. To avoid endless loops in case of unsatisfiable constraints in an update, the number of DSC steps can be supervised and bounded by an additional software component. When used as self-configuration algorithm in a running system, all computation and communication of DPA and DSC can be performed on lowest priority, to minimize the effect on running applications.

10.5 The Local Improvement Target

In this section we will introduce a metric for evaluation of constraint violations. We will show how this new metric can be calculated solely from local information from the DPA instances making it highly suitable for a distributed algorithm. Then we will outline a naïve distributed algorithm and demonstrate how such a DSC algorithm may lead to oscillations. In the following section, we will then provide a distributed algorithm, based on the same metric, which avoids this issue.

First, we define the local improvement target (LIT) of a task as a metric for the violation of that task's path latency constraints.

Definition 1 *Let the local improvement target* δ_{τ_i} *of task* $\tau_i \in \Gamma_{\rho_j}$ *be defined as*

$$\delta_{\tau_i} = \max_{\psi_k \in \Psi_{\tau_i}} \left(0, \frac{\omega_{\tau_i}}{\lambda_{\psi_k}} * (\lambda_{\psi_k} - \chi_{\psi_k})\right) \tag{10.1}$$

The LIT indicates the "responsibility" of a task for a path latency constraint violation. To obtain a task's LIT, the quotient of the task's WCRT and the path latency multiplied by the path violation is calculated for all paths, that the task is part of. The task's LIT is the maximum value obtained for this task. The calculation only requires the task's worst-case response times ω_{τ_i}, and path latencies λ_{ψ_k} and latency constraints χ_{ψ_k} of all paths $\psi_k \in \Psi_{\tau_i}$, that the task is part of. All of this information is provided by the DPA based on its local model. Thus, LITs can be calculated at DSC instances without the necessity of explicit communication among the different DSC instances.

A naïve algorithm to priority assignment is to assign priorities directly with ascending LITs, i.e. tasks with higher LITs receive higher priorities. This algorithm requires all DSC instances to concurrently calculate the LITs for all tasks on their respective resources. If any LIT is greater than 0, i.e. a constraint is violated, the corresponding DSC instance would sort the tasks in descending order of their respective LITs and assign priorities in descending order of that sorted list. This *direct LIT-based* approach proves feasible if applied to only one resource at a time as was shown with a similar metric in [1]. However, if applied to several resources concurrently, this algorithm may lead to oscillatory, as we show in the following.

Again, consider the example system shown in Fig. 10.2a. The system is composed of two resources connected over a bus and of two task sets (τ_1,τ_2 and τ_3,τ_4) with constraints on their end-to-end path latencies. For simplicity we shall disregard transmission times and scheduling on the bus. In this setup the path latency constraints may be assigned such that both constraints can only be satisfied if $\tau_1 > \tau_3$ and $\tau_2 < \tau_4$ or if $\tau_1 < \tau_3$ and $\tau_2 > \tau_4$ ($\tau_x > \tau_y$ denotes a higher priority for τ_x). This scenario is depicted in Fig. 10.2b. If in the initial state either τ_1 and τ_2 or τ_3 and τ_4 hold the higher priorities, the concurrent distributed priority assignment on both resources will result in oscillation. As the lower priority path will violate its constraint, both resources will give the corresponding tasks the higher priorities simultaneously, resulting in the inverted situation in the following DSC step. The

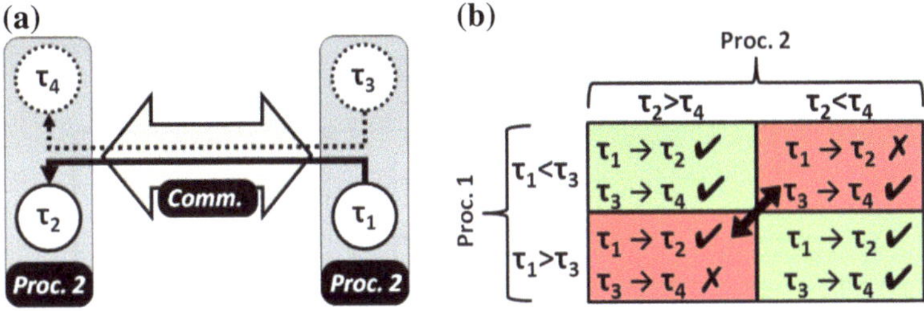

Fig. 10.2 Example system **a** example system. **b** possible configurations

other two states are the feasible solutions, that are never reached, i.e. the distributed algorithm gets stuck and alternates between the same infeasible configurations. As we will see in the evaluation (Sect. 10.7), such oscillations occur frequently if the direct LIT-based algorithm is used in a distributed setup, i.e. approaches as presented in [1] do not generalize to concurrent distributed setups. In the remainder of this chapter, we explain, how the problem of priority assignment can nonetheless be addressed with a highly efficient distributed algorithm.

10.6 Distributed Self-Configuration Algorithm

We propose to use a control-theory inspired approach within each DSC instance to assign priorities without causing oscillatory loops.

In the previous section we have seen, that oscillations may occur because DSC instances do not consider the behavior of other DSC instances in the priority assignment process. We propose to address this by considering past DSC steps in each instance. However, logging all evaluated configurations of previous DSC steps is intractable because 1. no DSC instance has a complete view of the system model and thus cannot decide alone whether a configuration has already been evaluated 2. logging all previous configurations introduces significant memory overhead, which may be prohibitive, if the algorithm is used in-system along with an admission control scheme.

We propose a DSC algorithm that complements the priority assignment in decreasing order of the LIT with a time-discrete PID filter, as depicted in Fig. 10.3. Such a filter allows to track past DSC steps with minimal memory overhead. In a first step (1. in Fig. 10.3) *set point priorities* S_{τ_i} are assigned in decreasing order of the tasks' LIT. This is the direct LIT-based priority assignment, which tends to oscillate if no further counter-measures are taken. These set point priorities are input to a feedback PID-controller (2. in Fig. 10.3). This filter returns a *priority rating* R_{τ_i} which incorporates the set-point priorities S_{τ_i}, the currently assigned priorities P_{τ_i} and the history of DSC steps through a proportional (P), an integral (I) and a derivative (D) component. The proportional component is equivalent to a scalable priority

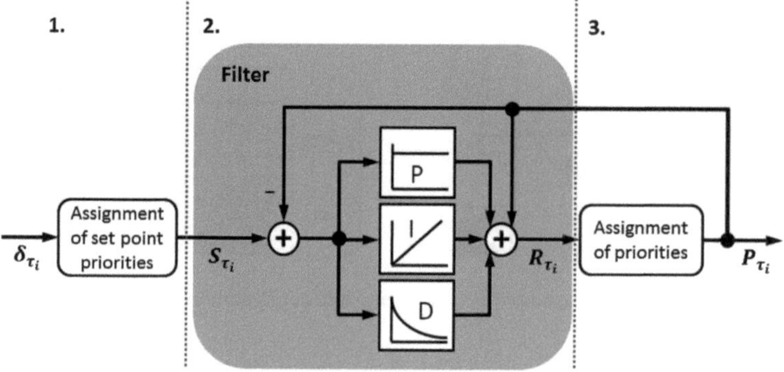

Fig. 10.3 Feedback control for priority assignment

assignment in direct correlation with the LIT. The integral component allows to super-proportionally increase the priority rating of a task if it violates any of its path latency constraints over several subsequent DSC steps. The derivative component dampens this effect by decreasing the priority rating if it increased in the previous DSC step. In combination all three components allow to calculate a sufficiently stable priority rating. Once the priority ratings have been calculated tasks are assigned priorities in decreasing order of these ratings (3. in Fig. 10.3).

The described flow is further detailed in algorithm 2, which shows the l-th DSC step. In a first step all local improvement targets are calculated (line 3). To assign the set point priorities for each task $\tau_i \in \Gamma_{\rho_j}$, Γ_{ρ_j} is sorted in descending order of the LITs δ_{τ_i} (line 6). The set point priority of a task τ_i is then set to its position in the sorted Γ_{ρ_j} (line 7).

The set point priorities S serve as input for the filter. Let $\Delta_{\tau_i}(l)$ be the difference of assigned priority and set point priority in the l-th DSC step, i.e.

$$\Delta_{\tau_i}(l) = S_{\tau_i}(l) - P_{\tau_i}(l-1) \tag{10.2}$$

The priority rating $R_{\tau_i}(l)$ in the l-th DSC step is then calculated by

$$\begin{aligned} R_{\tau_i}(l) = P_{\tau_i}&(l-1) \\ &+ k_P * \Delta_{\tau_i}(l) \\ &+ k_I * I_{\tau_i}(l) \\ &+ k_D * D_{\tau_i}(l) \end{aligned} \tag{10.3}$$

with

$$I_{\tau_i}(l) = I_{\tau_i}(l-1) + \Delta_{\tau_i}(l) \tag{10.4}$$

$$D_{\tau_i}(l) = \Delta_{\tau_i}(l-1) - \Delta_{\tau_i}(l) \tag{10.5}$$

Algorithm 1: l-th DSC Step (filtered LIT-based)

1: **for** $\rho_j \in \mathscr{P}_s \subseteq \mathscr{P}$ **concurrently do**
2: **for** $\tau_i \in \Gamma_{\rho_j}$ **do**
3: calculate δ_{τ_i}
4: **end for**
5: **if** any $\delta_{\tau_i} > 0 : \tau_i \in \Gamma_{\rho_j}$ **then**
6: sort Γ_{ρ_j} descending in δ_{τ_i}
7: assign set point priorities in order of sorted Γ_{ρ_j}
8: **for** $\tau_i \in \Gamma_{\rho_j}$ **do**
9: calculate $R_{\tau_i}(l)$
10: **end for**
11: sort Γ_{ρ_j} descending in $R_{\tau_i}(l)$
12: assign priorities in order of sorted Γ_{ρ_j}
13: **end if**
14: **end for**

and

$$k_P, k_I, k_D \in \mathbb{R} \tag{10.6}$$

The parameters k_P, k_I and k_D are the gain parameters of the proportional, integral and differential components of the filter, respectively. After calculation of the priority ratings $R_{\tau_i}(l)$ for all $\tau_i \in \Gamma_{\rho_j}$ (line 9), the set Γ_{ρ_j} is sorted in descending order of the priority ratings $R_{\tau_i}(l)$ (line 11). The priority of all tasks $\tau_i \in \Gamma_{\rho_j}$ is then set to their respective position in the sorted Γ_{ρ_j} (line 12).

Note that the addition of the filter does not prevent oscillations from ever occurring. The inputs of each DSC instance depend on the behavior of potentially several other DSC instances - while these dependencies change with different priority assignments. Thus, stability of the constraint solving process cannot be guaranteed. Furthermore, as the feedback path of the filter includes a sorting operation, it is intractable to calculate optimal gain parameters to achieve a certain damping. Instead we have determined suitable gain parameters (Eq. 10.7–10.9) empirically on the testcases that were used in the evaluation (Sect. 10.7).

In order to find suitable gain parameters we have observed the change of path latencies of several testcase systems over the course of several DSC steps. Fig. 10.4 shows a plot of this for one testcase system, which consists of 4 resources, 10 tasks and 5 communication channels on one communication medium. Three paths with constrained latency are defined. The plot shows the path latencies in solid lines and the respective constraint in the same color as dashed line. We see that the experimentally chosen gain parameters cause a decent settling behavior of the path latencies towards their respective constraints, rendering the system feasible after 5 DSC steps. The gain parameters that were used for this testcase as well as the evaluation in Section 10.7 are given below.

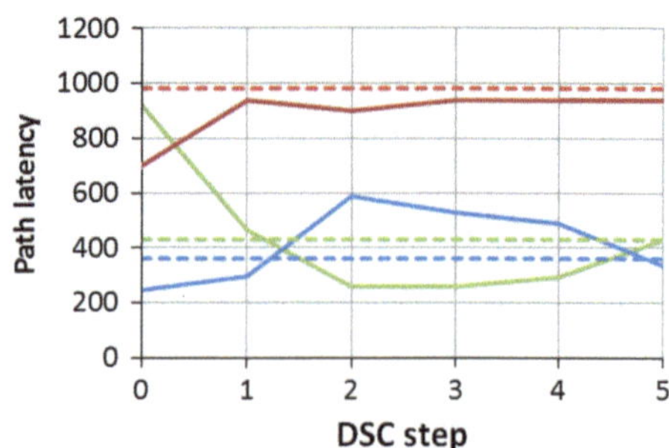

Fig. 10.4 Example: Path latency over several DSC steps

$$k_P = -0,4 \tag{10.7}$$

$$k_I = 0,05 \tag{10.8}$$

$$k_D = -0,1 \tag{10.9}$$

Although, the gain parameters define the settling behavior, their experimental determination has shown, that their influence is continuous and the distributed priority assignment algorithm is not overly sensitive to small changes in these parameters.

Furthermore, although we cannot guarantee optimality of the filter we observe, that the possibility to rank tasks based on their value-continuous priority rating—instead of the stepwise changing path latency—and because these priority ratings incorporate the history of DSC steps, a more stable trend for priority assignment is obtained. In the following section we demonstrate on a larger set of testcases that this filtering approach indeed poses a suitable approach to priority assignment.

10.7 Evaluation

In this section we evaluate the performance of the proposed algorithm. As it employs a heuristic, we have tested it on an extensive set of testcases. As baseline for the comparison we use a state of the art design-time tool, which is based on a genetic algorithm [12]. Furthermore, we compare the performance to the lazy algorithm presented in [20], which builds on the same general DSC approach as this chapter while reducing oscillations by means of a lazy threshold.

The testcase systems were generated with the open-source tool *System Models for Free* (SMFF) [28, 29], which pseudo-randomly generates completely specified system models. We have used two different parameter sets to evaluate scalability of the approach. The first parameter set generates smaller systems with 4 computational resources, 2–3 communication resources and 2–4 tasks per task set. The second parameter set generates system models with 12 computational resources, 3–5 communication resources and 3–7 tasks per task set. The number of task sets per testcase depends on the success of the filtered LIT-based algorithm and the GA.

	BatchPlatformFactory	resCnt	meanCResPercentage	*StdTgffApplicationFactory*	numTasks	diffNumTasks	tasMaxDegrIn	taskMaxDegrOut	*StdMapper*	kPredecessor	kSuccessor	kResDist	*RandomPriorityAssigner*	no parameters	*TaskLoadTimingFactory*	minActPeriod	maxActPeriod	minTaskLoad	maxTaskLoad	bcetPercentage	*LaxityConstraintFactor*	minConstrFactor	maxConstrFactor
Test 1		3	0,35		3	1	1	1		3,0	3,0	1,5		n.a.		1000	2000	1%	5%	1,0		3,0	5,0
Test 2		12	0,35		5	2	1	1		3,0	3,0	1,5		n.a.		1000	2000	1%	5%	1,0		3,0	5,0

Fig. 10.5 Parameters for testcase generation

We have added additional task sets until the system turned infeasible. Then we have used both algorithms to find a feasible configuration. If one succeeded, we added another task set. This was repeated until neither algorithm was able to find a feasible configuration. As a result the different testcases contained between 2 and 10 task sets (i.e. in total 4–70 tasks) and in total 118 testcases were analyzed per parameter set. For both parameter sets the worst-case execution times of each task were set such that it caused a load (i.e. WCET/Period) between 1 and 5 %. Constraints on end-to-end path latency were set to values 3–5 times larger than the sum of the WCETs of all tasks along the path. For reproducibility of results we provide the complete set of parameters in Fig. 10.5.

10.7.1 Number of Feasible Priority Assignments

First, we regard the number of testcase systems, for which the LIT-based algorithm could find a feasible priority assignment. In many cases it is intractable, to analyze whether a system has a feasible priority assignment at all, as the number of possible configurations easily reaches values greater than several 100 million. As a consequence we perform this evaluation relative to the genetic algorithm of [12].

We will call an optimization run a *false negative* for the filtered LIT-based algorithm/GA, if the filtered algorithm/GA failed to find a solution, while the other did find a feasible priority assignment, respectively. The LIT-based algorithm was restricted to 500 DCS steps, while the GA analyzed 50 individuals over 10 generations. Fig. 10.6a shows the percentage of testcases where only the filtered LIT-based algorithm (green), both algorithms (yellow) and only the GA (red) found a solution. For the small systems the filtered LIT-based algorithm and the genetic algorithm [12] performed equally well. In more than 80 % of the testcases both algorithms found a solution. In ~8 % of the testsystems each algorithm was able to find a solution while the other failed to find a feasible priority assignment. From the larger testcase

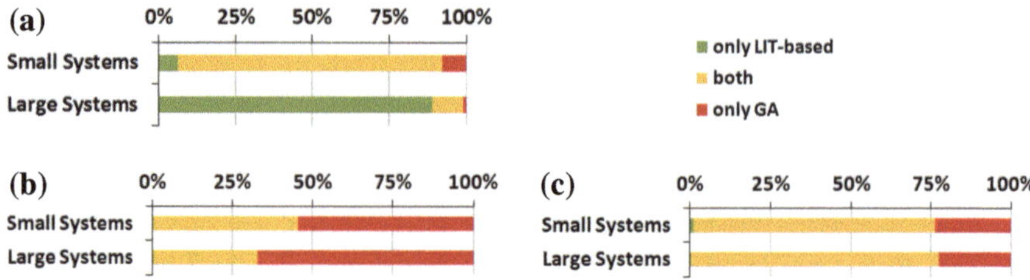

Fig. 10.6 Comparison w.r.t. solved testcases **a** filtered LIT-based **b** direct LIT-based [20] **c** 15 %-lazy LIT-based [20]

systems we see that the filtered algorithm scales significantly better than the GA. In ~90 % of the large testcase systems only the filtered algorithm was able to find a feasible priority assignment while the GA failed to find a solution. Conversely, in only ~1 % of the larger testcases the GA found a solution while our novel approach failed to solve the priority assignment problem.

As we have used the same test setup as [20] we can also compare the performance of the filtered algorithm to the direct LIT-based algorithm (Fig. 10.6b) and to the lazy LIT-based algorithm [20] (Fig. 10.6c). We see that both previous LIT-based approaches, unlike the proposed filtered algorithm, were outperformed by the GA.

10.7.2 Runtime

Next we compare the runtime of the algorithms. Because the runtime of the self-configuration algorithm is dominated by the underlying DPA, and the benchmark GA uses the same performance analysis, we use the number of required performance analysis runs (i.e. number of DSC steps) to obtain a solution as runtime metric. We introduce the comparison factor c, that signifies the relation of the number of analysis runs needed by the GA b and the number of analysis runs needed by the distributed LIT-based algorithm a.

$$c = \begin{cases} b/a & \text{if } a < b \\ 0 & \text{if } a = b \\ -a/b & \text{if } b < a \end{cases} \tag{10.10}$$

For both parameter sets Fig. 10.7 shows histograms over the comparison factor c for those testcases where both algorithms found a solution. Regardless of the testcase size the proposed LIT-based algorithm requires an order of magnitude less performance analyses to derive a feasible priority assignment than the GA, which is a state of the art design time tool (12.75× and 11.48× average improvement for the small and large testcase systems, respectively). For the same set of testcase generation parameters the lazy LIT-based algorithm of [20] states average runtime improvements vs.

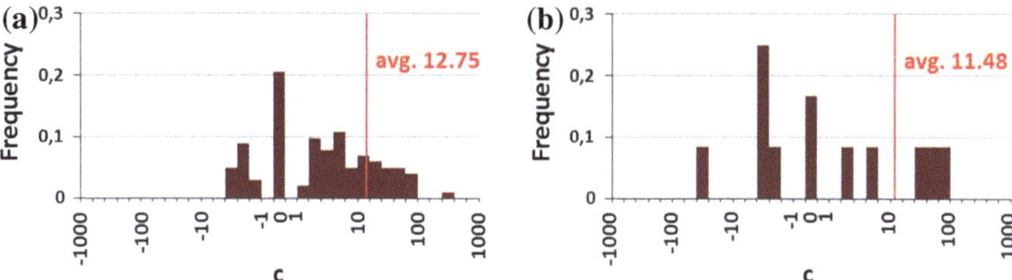

Fig. 10.7 Improvement histograms **a** filtered LIT-based, Param. Set 1 **b** filtered LIT-based, Param. Set 2

the GA of $5\times$ and $7\times$ for the two testcase parameter set. Thus, the presented filtered algorithm is $\sim 2\times$ faster than the lazy algorithm while it is able to solve significantly more testcases.

10.8 Conclusion

In this chapter we have presented an algorithm that distributedly finds feasible priority assignments in distributed SPP scheduled systems under consideration of end-to-end path latency constraints. Due to the possibility of distributed implementation the algorithm can be used to complement an admission control scheme as in [2] to enhance a system by self-configuration capabilities.

In an evaluation based on an extensive set of pseudo-randomly generated testcases we have shown that the proposed filtered LIT-based algorithm was able to solve as many small testcases systems as a current design-time tool. For larger more complex systems it even outperformed the existing software significantly. At the same time, irrespective of the testcase size the proposed algorithm was more than an order of magnitude faster than the design-time tool.

Thus, although designed for an in-system distributed implementation, the algorithm poses an attractive choice for design-time configuration synthesis.

References

1. Stein, S., Hamann, A., Ernst, R.: Real-time property verification in organic computing systems. In: Second Int'l. Symp. on Leveraging Applications of Formal Methods, Verification and Validation (2006)
2. Neukirchner, M., Stein, S., Schrom, H., Ernst, R.: A software Update Service with Self-Protection Capabilities. In: Conf. on Design, Automation and Test in Europe (DATE) (2010)
3. Henia, R., Hamann, A., Jersak, M., Racu, R., Richter, K., Ernst, R.: System level performance analysis - the SymTA/S approach. Computers and Digital Techniques, IEE Proc. - 152, 148–166 (2005).doi:10.1049/ip-cdt:20045088

4. Liu, C.L., Layland, J.W.: Scheduling algorithms for multiprogramming in a hard-real-time environment. J. ACM **20**, 46–61 (1973)
5. Leung, J.Y.T., Whitehead, J.: On the complexity of fixed-priority scheduling of periodic, real-time tasks. Perform. Eval. **2**, 237–250 (1982)
6. Audsley, N., Burns, A., Richardson, M.F., Wellings, A.J.: Hard real-time scheduling: The deadline-monotonic approach. In: proceeding IEEE Workshop on Real-Time Operating Systems and Software (1991)
7. Lehoczky, J., Ramos-Thuel, S.: An optimal algorithm for scheduling soft-aperiodic tasks in fixed-priority preemptive systems. Real-Time Systems Symposium (RTSS), pp. 110–123 (1992)
8. Davis, R., Burns, A.: Optimal priority assignment for aperiodic tasks with firm deadlines in fixed priority pre-emptive systems. Inf. Process. Lett. **53**, 249–254 (1995)
9. Bertogna, M., Cirinei, M., Lipari, G.: New schedulability tests for real-time task sets scheduled by deadline monotonic on multiprocessors. Principles of Distributed Systems. Springer, Heidelberg, (2006)
10. Andersson, B.: Global static-priority preemptive multiprocessor scheduling with utilization bound 38%. Principles of Distributed Systems. Springer, Heidelberg (2008)
11. Davis, R.I., Burns, A.: Priority assignment for global fixed priority pre-emptive scheduling in multiprocessor real-time systems. In: Real-Time Systems Symposium (RTSS) (2009)
12. Hamann, A., Jersak, M., Richter, K., Ernst, R.: A framework for modular analysis and exploration of heterogeneous embedded systems. Real-Time Syst. **33**, 101–137 (2006)
13. Glaß, M., Lukasiewycz, M., Teich, J., Bordoloi, U., Chakraborty, S.: Designing heterogeneous ECU networks via compact architecture encoding and hybrid timing analysis. In: Proceedings of Design Automation Conference (DAC), pp. 43–46 (2009)
14. Palopoli, L., Abeni, L., Cucinotta, T., Lipari, G., Baruah, S.: Weighted feedback reclaiming for multimedia applications. In: Workshop on Embedded Systems for Real-Time Multimedia (ESTImedia) (2008). doi:10.1109/ESTMED.2008.4697009
15. Cucinotta, T., Palopoli, L.: QoS control for pipelines of tasks using multiple resources. IEEE Trans. Comput. **59**, 416–430 (2010)
16. Jonsson, J., Shin, K.G.: Robust adaptive metrics for deadline assignment in distributed hard real-time systems. Real-Time Syst. **23**, 239–271 (2002)
17. García, J.G., Harbour, M.G.: Optimized priority assignment for tasks and messages in distributed hard real-time systems. Workshop on Parallel and Distributed Real-Time Systems, (1995)
18. Natale, M.D., Stankovic, J.A.: Dynamic end-to-end guarantees in distributed real time systems. In: Real-Time Systems Symp. (RTSS) (1994)
19. Hong, S., Chantem, T., Hu, X.S.: Meeting end-to-end deadlines through distributed local deadline assignments. In: Real-Time Systems Symposium (RTSS) (2011)
20. Neukirchner, M., Stein, S., Ernst, R.: A lazy algorithm for distributed priority assignment in real-time systems. Workshop on Self-Organizing Real-Time Systems (SORT), (2011)
21. Richter, K.: Compositional scheduling analysis using standard event models. Ph.D. thesis, Technical University of Braunschweig, Department of Electrical Engineering and Information Technology (2004)
22. Thiele, L., Chakraborty, S., Naedele, M.: Real-time calculus for scheduling hard real-time systems. In: Symposium on Circuits and Systems (ISCAS) (2000). doi:10.1109/ISCAS.2000.858698
23. Tindell, K.W.: An extendible approach for analysing fixed priority hard real-time systems. J. Real-Time Syst. **6**, 133–152 (1994)
24. Racu, R., Li, L., Henia, R., Hamann, A., Ernst, R.: Improved response time analysis of tasks scheduled under preemptive round-robin. Conference on Hardware-Software Codesign and System Synthesis, (2007)
25. Davis, R.I., Burns, A., Bril, R.J., Lukkien, J.J.: Controller area network (can) schedulability analysis: Refuted, revisited and revised. Real-Time Syst. **35**, 239–272 (2007)

26. Schliecker, S., Ernst, R.: A recursive approach to end-to-end path latency computation in heterogeneous multiprocessor systems. In: Conference on Hardware Software Codesign and System Synthesis (CODES-ISSS) (2009)
27. Stein, S., Neukirchner, M., Schrom, H., Ernst, R.: Consistency challenges in self-organizing distributed hard real-time systems. Workshop on Self-Organizing Real-Time Systems (SORT), (2010)
28. Neukirchner, M., Stein, S., Ernst, R.: SMFF: System Models for Free. In: 2nd Int'l. Workshop on Analysis Tools and Methodologies for Embedded and Real-time Systems (WATERS) (2011)
29. Neukirchner, M.: System models for free. http://smff.sourceforge.net (2011). http://smff.sourceforge.net

Chapter 11
Exploration of Distributed Automotive Systems Using Compositional Timing Analysis

Martin Lukasiewycz, Michael Glaß, Jürgen Teich and Samarjit Chakraborty

Abstract This chapter presents a design space exploration method for mixed event-triggered and time-triggered real-time systems in the automotive domain. A design space exploration model is used that is capable of modeling and optimizing state-of-the-art automotive systems including the resource allocation, task distribution, message routing, and scheduling. The optimization is based on a heuristic approach that iteratively improves the system design. Within this iterative optimization it is necessary to analyze each system design where one of the major design objectives that needs to be evaluated is the timing behavior. Since timing analysis is a very complex design task with high computational demands, it might become a bottleneck within the design space exploration. As a remedy, a clustering strategy is presented that is capable of reducing the complexity and minimizing the runtime of the timing analysis. A case study gives evidence of the efficiency of the proposed approach.

11.1 Introduction

Automotive electronics are constantly becoming more complex due to the innovation pressure in the automotive domain. A vast majority of innovations in the automotive domain is nowadays driven by embedded systems. In the last years such innovations

M. Lukasiewycz (✉)
TUM CREATE, Singapore, Singapore
e-mail: martin.lukasiewycz@tum-create.edu.sg

M. Glaß · J. Teich
Friedrich-Alexander-Universität Erlangen-Nürnberg, Erlangen, Germany
e-mail: glass@cs.fau.de

J. Teich
e-mail: teich@cs.fau.de

S. Chakraorty
TU Munich, Munich, Germany
e-mail: samarjit@tum.de

A. Sangiovanni-Vincentelli et al. (eds.), *Embedded Systems Development*, Embedded Systems 20, DOI: 10.1007/978-1-4614-3879-3_11,

were for example adaptive cruise control, pedestrian detection, or intelligent parking assist systems. However, these innovations require increasingly sophisticated system architectures. As a result, top-of-the-range vehicles already contain up to 100 Electronic Control Unit (ECU) and a multitude of different bus systems. In case functions have stringent latency and jitter constraints, these systems often require a complex validation of end-to-end timing behavior, using tools like Sympolic Timing Analysis for Systems (SymTA/S) [1] or Modular Performance Analysis (MPA) [2, 3]. This evaluation is a challenging design task and might become a bottleneck within a Design Space Exploration (DSE) where an optimization of the resource allocation, task mapping, message routing, and scheduling is performed. As a remedy, this chapter presents a (DSE) approach that uses efficient timing analysis based on a graph-based representation and a fine-grained fixed-point iteration that partitions the problem in case of cyclic dependencies. In the following, the DSE model is introduced. Based on this model, an approach is presented that is capable of reducing the runtime of the timing analysis significantly. This is done by a decomposition of the timing analysis problem and an ordered evaluation. Finally, a case study is presented that gives evidence of the efficiency of the proposed approach.

11.2 Design Space Exploration Model

In the following, the Design Space Exploration (DSE) model is introduced, see [4, 5]. It is based on the optimization approach presented in [6]. This optimization approach is based on an Evolutionary Algorithm (EA), supporting multiple and non-linear objectives. For this purpose it becomes necessary to define the model formally and encode it into a set of linear constraints with binary variables such that a feasible implementation corresponds to a feasible solution. The remaining optimization, including load balancing and non-linear constraint satisfaction, is automatically carried out in an iterative search process of the optimization approach. This procedure significantly reduces the efforts to implement a new optimization approach or define complex constraints to direct the search towards the optimal implementations.

11.2.1 Model Description

The used exploration model is defined by a *specification* that consists of an *application* and an *architecture*. *Mappings* represent the relation between the process tasks from the application to the architecture, indicating which process task can be implemented on which resource. The application model also supports multi-cast and multi-hop communication by introducing messages additionally to the process tasks. A message can be routed on every resource except those where a routing is explicitly prohibited. From this specification, various *implementations* are derived. The implementation is defined by the *allocation* of architecture resources from a set

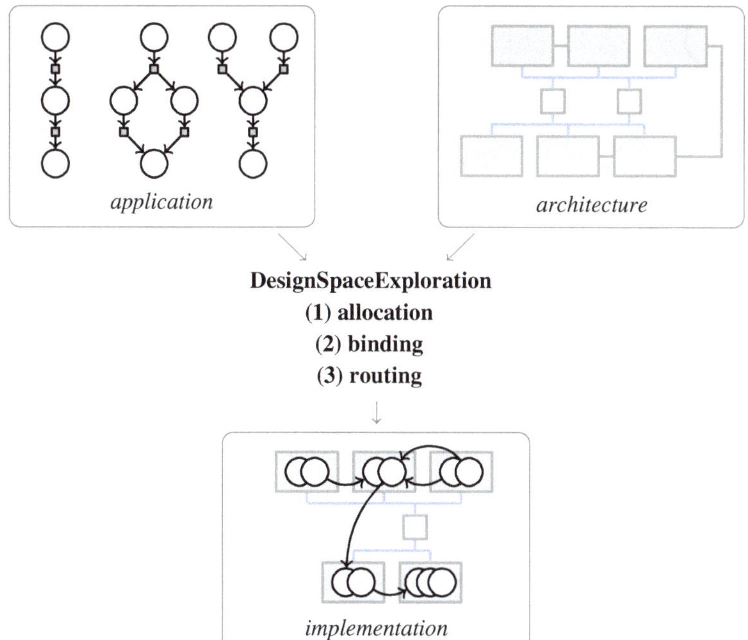

Fig. 11.1 Illustration of the Y-chart approach for the Design Space Exploration (*DSE*) model. An application is mapped to an architecture, resulting in an implementation. The DSE performs an allocation of resources, the binding of process tasks, and the routing of messages

of predefined components, the *binding* of process tasks to resources and *routing* of messages. The Y-chart approach for this model is illustrated in Fig. 11.1.

The specification consists of an architecture graph G_R, an application graph G_T, and mapping edges E_M:

- The architecture is given by a directed graph $G_R(R, E_R)$. The vertices R represent resources such as ECUs, gateways, and bus systems. The directed edges $E_R \subseteq R \times R$ indicate available communication connections between resources.
- The application is given by a directed graph $G_T(T, E_T)$ with $T = P \cup C$. The vertices T are either process tasks $p \in P$ or messages $c \in C$. Each edge $e \in E_T$ connects a vertex in P to one in C, or vice versa. Each process task can have multiple incoming edges that indicate the data dependencies to communication information of the predecessor messages. A process task can also have multiple outgoing edges to allow the sending of multiple different messages. On the other hand, each message has exactly one predecessor process task as the sender. To allow multi-cast communication, each message can have multiple successor process tasks.
- The set of mapping edges E_M contains the mapping information for the process tasks. Each mapping edge $m = (p, r) \in E_M$ indicates a possible implementation

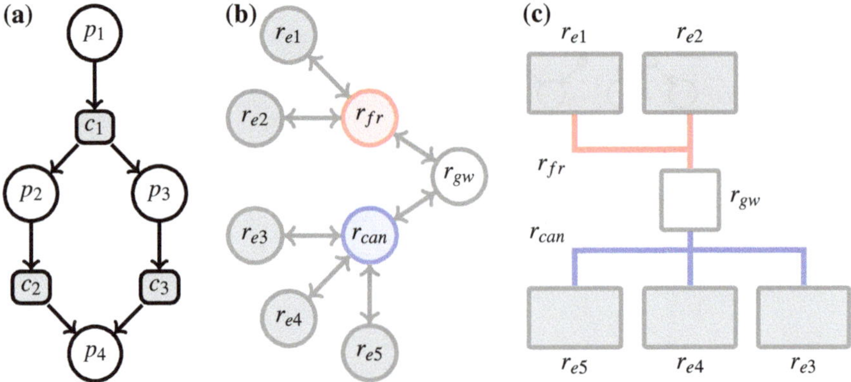

Fig. 11.2 Specification with the application graph G_T (**a**) and architecture graph G_R (**b**) for a given architecture (**c**). The mapping edges are defined as follows: $E_M = \{(p_1, r_{e1}), (p_1, r_{e2}), (p_2, r_{e2}), (p_3, r_{e2}), (p_3, r_{e3}), (p_3, r_{e4}), (p_4, r_{e5})\}$

of the process $p \in P$ on the resource $r \in R$. Without loss of generality it is assumed that messages can be routed on every resource.

A sample specification is given in Fig. 11.2. This specification comprises a Control Area Network (CAN) bus (r_{can}), a FlexRay bus (r_{fr}), and a gateway (r_{gw}) that interconnects the buses. The communication over the buses and the gateway can only be established by multiple hops.

One implementation consists of the allocation graph G_α that is deduced from the architecture graph G_R and the binding E_β as a subset of E_M that maps the application to the allocation. Additionally, for each message $c \in C$ a sub-graph of the allocation $G_{\gamma,c}$ is determined that fulfills the data dependencies such that the communication is established between each sender process task and the corresponding receiver process tasks.

- The allocation is a directed graph $G_\alpha(\alpha, E_\alpha)$ that is an induced sub-graph of the architecture graph G_R. The allocation contains all resources that are available in the current implementation. The edges are induced from the graph G_R such that G_α is aware of all communication connections.
- The binding is performed by a mapping of the tasks to the allocated resources by deducing E_β from E_M such that the following requirements are fulfilled. Each process task $p \in P$ in the application is bound to exactly one resource:

$$\forall p \in P : |\{m | m = (p, r) \in E_\beta\}| = 1 \tag{11.1}$$

Each task can only be bound to allocated resources:

$$\forall m = (p, r) \in E_\beta : r \in \alpha \tag{11.2}$$

- Each message in $c \in C$ is routed on a tree $G_{\gamma,c}$ that is a sub-graph of the allocation G_α. The routings have to be performed such that all data dependencies given by the following two conditions are satisfied.
  For each message $c \in C$, the root of the routing has to equal the binding resource of the predecessor sender process task $p \in P$:

$$\forall (p, c) \in E_T, m = (p, r) \in E_\beta : |\{e | e = (\tilde{r}, r) \in G_{\gamma,c}\}| = 0 \tag{11.3}$$

  Each message $c \in C$ has to be routed on the same resource as the binding resource of the successive process tasks $p \in P$:

$$\forall (c, p) \in E_T, m = (p, r) \in E_\beta : r \in G_{\gamma,c} \tag{11.4}$$

An implementation is *feasible* if all requirements on the process task binding and the routing of the messages, i. e., the data dependencies, are fulfilled. A feasible implementation for the specification in Fig. 11.2 is given in Fig. 11.3.

11.2.2 Binary Encoding

In the following, a set of linear constraints with binary variables is defined such that a solution $\mathbf{x} \in \{0, 1\}^n$ corresponds to a *feasible* implementation x for the given DSE problem. The symbolic encoding uses the following binary variables:

r–binary variable for each resource $r \in R$ indicating whether this resource is in the allocation α (1) or not (0)

m–binary variable for each mapping $m \in E_M$ indicating whether the mapping edge is in E_β (1) or not (0)

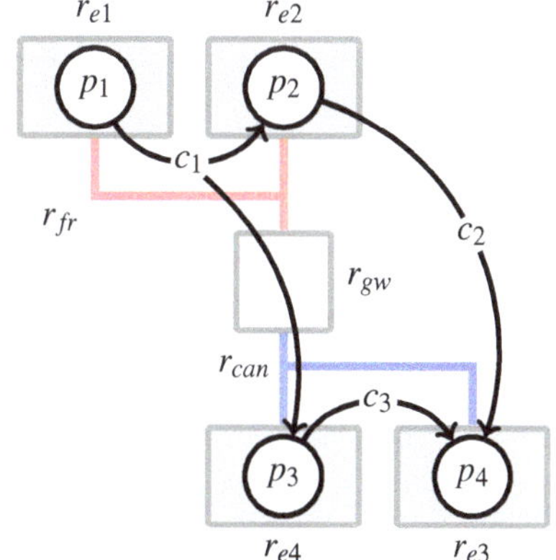

Fig. 11.3 Implementation for the specification in Fig. 11.2. Illustrated is the allocation G_α, binding E_β, and routing G_γ. All routings are performed within multiple hops using the available buses and the gateway

$\mathbf{c_r}$–binary variable for each message $c \in C$ and the available resources $r \in R$ indicating whether the message is routed on the resource (1) or not (0)
$\mathbf{c_{r,t}}$–binary variable for each message $c \in C$ and resource $r \in R$ indicating on which communication step $t \in \mathcal{T} = \{1, .., |\mathcal{T}|\}$ (messages are propagated in steps or hops, respectively) a message is routed on the resource

The linear constraints are formulated as follows:
$\forall p \in P$:

$$\sum_{m=(p,r)\in E_M} \mathbf{m} = 1 \tag{11.5}$$

$\forall m = (p, r) \in E_M$:

$$\mathbf{r} - \mathbf{m} \geq 0 \tag{11.6}$$

$\forall c \in C, r \in R, (c, p) \in E_T, m = (p, r) \in E_M$:

$$\mathbf{c_r} - \mathbf{m} \geq 0 \tag{11.7}$$

$\forall c \in C$:

$$\sum_{r\in R} \mathbf{c_{r,1}} = 1 \tag{11.8}$$

$\forall c \in C, r \in R, (p, c) \in E_T, m = (p, r) \in E_M$:

$$\mathbf{m} - \mathbf{c_{r,1}} = 0 \tag{11.9}$$

$\forall c \in C, r \in R$:

$$\sum_{t\in\mathcal{T}} \mathbf{c_{r,t}} \leq 1 \tag{11.10}$$

$\forall c \in C, r \in R$:

$$\left(\sum_{t\in\mathcal{T}} \mathbf{c_{r,t}}\right) - \mathbf{c_r} \geq 0 \tag{11.11}$$

$\forall c \in C, r \in R, t \in \mathcal{T}$:

$$\mathbf{c_r} - \mathbf{c_{r,t}} \geq 0 \tag{11.12}$$

$\forall c \in C, r \in R, t = \{1, .., |\mathcal{T}|\}$:

$$\left(\sum_{\tilde{r}\in R, e=(\tilde{r},r)\in E_R} \mathbf{c_{\tilde{r},t}}\right) - \mathbf{c_{r,t+1}} \geq 0 \tag{11.13}$$

$\forall c \in C, r \in R$:

$$\mathbf{r} - \mathbf{c_r} \geq 0 \tag{11.14}$$

$\forall r \in R:$

$$\left(\sum_{c \in C \wedge r \in R} \mathbf{c_r}\right) + \left(\sum_{m=(p,r) \in E_M} \mathbf{m}\right) - \mathbf{r} \geq 0 \tag{11.15}$$

The constraints in Eqs. (11.5) and (11.6) fulfill the binding of each task to exactly one resource and the requirement that tasks are only bound to allocated resources, respectively, as stated in Eqs. 11.1 and 11.2. A message has to be routed on each target resource of the successive process task mapping targets as stated in the requirement in Eq. (11.4). This requirement is fulfilled by the constraints in Eq. (11.7). Analogously, as stated in the requirement in Eq. (11.3), the constraints in Eqs. (11.8) and (11.9) imply that each message has exactly one root that equals the target resource of the predecessor mapping. The constraints in Eq. (11.10) ensure that a message can pass a resource at most once such that cycles are prohibited. A message has to be existent in one communication step on a resource in order to be correctly routed on this resource as implied by the constraint in Eqs. (11.11) and (11.12). The constraints in Eq. (11.13) state that a message may be routed only between adjacent resources in one communication step. In order ensure that the routing of each message is a sub-graph of the allocation, each message can be only routed on allocated resources as stated in the constraints in Eq. (11.14). Additionally, the constraints in Eq. (11.15) ensure that a resource is only allocated if it is used by at least one process or message such that suboptimal implementations are removed effectively from the search space. This minimizes the resulting allocation by redundant resources such that additional unnecessary costs are prohibited.

Given a single solution $\mathbf{x}$ of the defined set of linear constraints, a corresponding implementation x may be deduced as follows: The allocation G_α is deduced from the variables $\mathbf{r}$ and the binding E_β from the variables $\mathbf{m}$. For each message $c \in C$, the routing $G_{\gamma,c}$ is deduced from the variables $\mathbf{c_r}$ and $\mathbf{c_{r,t}}$.

11.3 Compositional Timing Analysis

The previous section defines a DSE model that is used to obtain feasible implementations x. Additionally, the exploration may also define priorities and schedules for the tasks and messages, respectively. For each implementation x, a timing analysis has to be performed to discard implementations that do not fulfill the real-time constraints of applications. In the following, a compositional timing analysis is proposed that is capable of determining end-to-end latencies efficiently in case of cyclic dependencies. Note that this approach reduces the runtime significantly without introducing any errors or additional over-approximations in the results.

11.3.1 Timing Model

In the used model it is assumed that the Worst-Case Execution Times (WCET) of all tasks and the transmission times of messages are known. Also the periods of applications are predefined and the priorities of tasks and messages are either predefined or determined by the DSE.

The proposed compositional timing analysis approach may take advantage of different analysis techniques. For example, Modular Performance Analysis (MPA) [3] is used for modeling the FlexRay bus protocol. For analyzing the Control Area Network (CAN) bus, the approach presented by Tindell et al. [7] is applied. Here, it may be noted that these approaches have different mechanisms for representing timing properties of message streams. The approach in [7] uses the traditional period and jitter event model. On the other hand, MPA uses a more generic event model based on arrival curves. Further details on arrival curves may be found in [3]. A method to convert arrival curves into standard event models and vice versa is presented in [8]. Using this method, the proposed compositional timing analysis enables a hybrid approach such that standard event models like *periodic*, *periodic with jitter*, and *sporadic* might be used as well as arbitrary arrival patterns represented by appropriate arrival curves.

Though the timing analysis may be performed efficiently with the described models, cyclic dependencies in the timing analysis require a fixed-point iteration that may become computationally expensive. Due to the dependencies, the timing properties have to be calculated iteratively until there are no changes anymore. For this purpose, different approaches for an efficient fixed-point iteration for timing analysis are proposed. The model requires a graph-based representation of timing dependencies where the basic element is the *timing entity*. A timing entity might, for example, be the execution of a process on an ECU or a transmission of a message on a bus or gateway. The goal of the timing analysis is to determine the *timing properties* for each timing entity within a compositional approach, i. e., separately from other calculations. The timing properties are usually a delay and jitter where the jitter might be a single real value or an arrival curve known from MPA.

In the following, a common global dependency-based fixed-point iteration as well as the proposed fine-grained fixed-point iteration approach are presented. The automotive network in Fig. 11.4 is introduced as an example to illustrate the proposed approaches.

11.3.2 Dependency-Based Fixed-Point Iteration

The determination of the timing properties of a timing entity might depend on the timing properties of other entities. It is suggested to use a graph-based representation with $G_\chi(V_\chi, E_\chi)$ where V_χ is the set of timing entities and E_χ a set of directed edges that define the dependency between timing entities. An edge $(v, \tilde{v}) \in E_\chi$

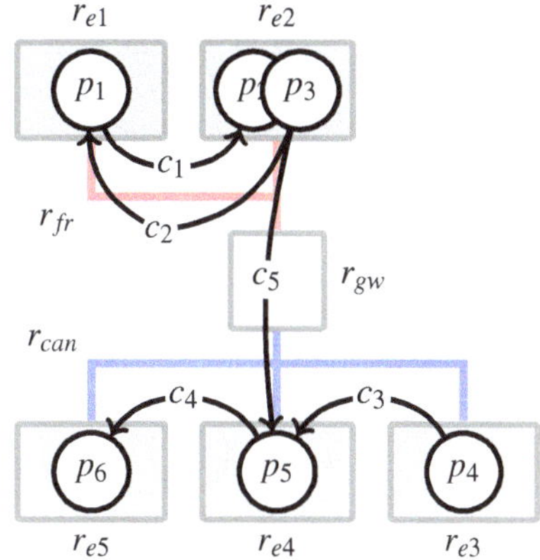

Fig. 11.4 A small automotive network consisting of five ECUs ($\{r_{e1}, ..., r_{e5}\}$), a CAN bus (r_{can}), a FlexRay bus (r_{fr}), and a gateway (r_{gw}). For the FlexRay bus, the static segment is used. The function consists of six processes ($\{p_1, ..., p_6\}$) communicating via five messages ($\{c_1, ..., c_5\}$). The index of the processes and messages represents the priority: A small number implies a high priority

indicates that the determination of the timing properties for $\tilde{v}$ depends on the timing properties of v. If such a *dependency graph* G_χ is acyclic, the timing properties may be determined in the partial order of the graph defined by the directed edges. In case of cycles in the graph, a *fixed-point iteration* becomes necessary to determine the timing properties of all entities. The dependency graph for the automotive network in Fig. 11.4 is given in Fig. 11.5 (a):

- All tasks or messages have an influence on the successive tasks or messages, respectively.
- Each process task on an ECU may delay the lower priority tasks on the same ECU (p_2 and p_3).
- Each message on the CAN bus may delay lower priority messages on the same CAN bus (c_3, c_4, and c_5).
- All messages on the FlexRay bus do not influence each other directly since they are routed on the static segment using Time Division Multiple Access (TDMA) (c_1 and c_2).

An algorithm for a dependency-based fixed-point iteration is given in Algorithm 1: The algorithm determines a fixed point for a subset $V \subseteq V_\chi$ of all timing entities. The set V_a contains all timing entities that shall be evaluated, i. e., starting with V (line 1). The iterative algorithm proceeds until the set V_a is empty (line 2). In each iteration, one element from V_a is selected and removed (line 3, 4). If the determined timing properties for v do not equal the previous value (line 5), all direct successive timing entities in G_χ that are also in V have to be re-evaluated (line 6). Note that the initial timing $t_0(v)$ for each entity, i. e., the initial jitter and delay, is 0 or is determined by initial timing approximations like presented in [9, 10].

Applying the Algorithm 1 to all timing entities, i. e., $V = V_\chi$, results a global dependency-based fixed-point iteration. Apparently, this approach is more efficient than a plain global fixed-point iteration that calculates the timing properties for all

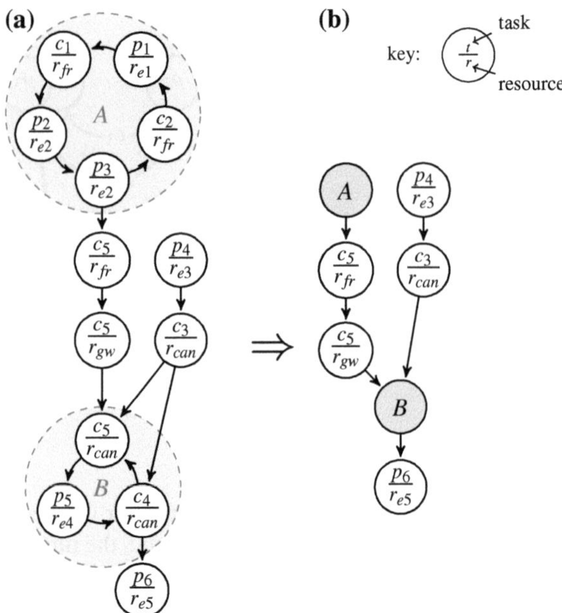

Fig. 11.5 Dependency graph G_χ (**a**) and acyclic dependency graph G_ψ (**b**) with merged states $A = \{(p_1, r_{e1}), (p_2, r_{e2}), (p_3, r_{e2}), (c_1, r_{fr}), (c_2, r_{fr})\}$ and $B = \{(p_5, r_{e4}), (c_4, r_{can}), (c_5, r_{can})\}$ for the automotive network in Fig. 11.4

Algorithm 1: Dependency-based fixed-point iteration that is applied on a subset $V \subseteq V_\chi$ of timing entities.

$V_a = V$;
while $V_a \neq \{\}$ **do**
  $v \in V_a$;
  $V_a = V_a \setminus \{v\}$;
  if $t_i(v) \neq t_{i-1}(v)$ **then**
    $V_a = V_a \cup (\{\tilde{v} | (v, \tilde{v}) \in E_\chi\} \cap V)$;
  end
end

timing entities in each iteration until no value is changed. However, as shown in the following, this approach can be further improved by a fine-grained approach.

11.3.3 Fine-Grained Fixed-Point Iteration

In the the following, a fixed-point iteration approach that is based on the stepwise calculation of timing entities is proposed. For the fine-grained fixed-point iteration, an acyclic graph $G_\psi(V_\psi, E_\psi)$ is deduced from G_χ. The vertices of the graph G_ψ are subsets of timing entities such that for each subset $V \in V_\psi$ it holds $V \subseteq V_\chi$. Furthermore, each timing entity is included in exactly one vertex:

$$\bigcup_{V \in V_\psi} V = V_\chi \tag{11.16}$$

$$\forall\, V, \tilde{V} \in V_\psi \text{ with } V \neq \tilde{V} : V \cap \tilde{V} = \{\} \tag{11.17}$$

Based on the partial order in G_ψ, a fixed-point iteration is applied to each node $V \in V_\psi$, i.e., the set of timing entities in V. These local fixed-point iterations are performed with the efficient dependency-based fixed-point iteration approach in Algorithm 1. Thus, a potentially inefficient global fixed-point iteration over all timing entities in V_χ is avoided. The acyclic dependency graph for the automotive network in Fig. 11.4 is given in Fig. 11.5 (b).

Given a dependency graph G_χ, an acyclic dependency graph G_ψ fulfilling the requirements in Eqs. (11.16) and (11.17) might be derived. To enable the best possible benefit from the introduced fine-grained fixed-point iteration, the optimal graph G_ψ shall contain a maximal number of vertices. A high number of vertices in G_ψ results in a high number of separate fine-grained fixed-point iteration steps and, thus, a more efficient approach is enabled.

In order to define the optimal graph G_ψ, the following reachability analysis is required. The function $r : V_\chi \rightarrow 2^{V_\chi}$ determines the set of reachable nodes in the graph G_χ from a node v and is defined recursively as follows:

$$r(v) = \{\tilde{v} \cup r(\tilde{v}) | (v, \tilde{v}) \in E_\chi\} \tag{11.18}$$

In a correct and optimal graph G_ψ, all timing entities with the same reachability are merged in the same vertex:

$$(\exists V \in V_\psi : v, \tilde{v} \in V) \Leftrightarrow r(v) = r(\tilde{v}) \tag{11.19}$$

The graph G_ψ is defined as *optimal* if only vertices with the same reachability are merged, i.e., the number of vertices in G_ψ is maximal. On the other hand, Eq. (11.19) ensures that a graph G_ψ contains no cycles, i.e., it is *correct*.

An efficient approach that merges the vertices with the same reachability from G_χ to the vertices in G_ψ can be done in $O(n^2)$ with $n = |V_\psi|$ as presented in the following: Correspondingly to the forward-reachability from Eq. (11.18), the backward-reachability is defined by $\overline{r} : V_\chi \rightarrow 2^{V_\chi}$ in the following recursive formulation:

$$\overline{r}(v) = \{\tilde{v} \cup \overline{r}(\tilde{v}) | (\tilde{v}, v) \in E_\chi\} \tag{11.20}$$

For any $v \in V_\chi$, the operation

$$V = r(v) \cap \overline{r}(v) \tag{11.21}$$

determines a vertex $V \in V_\psi$ containing all vertices on a cycle that contains v in G_χ. As a result, all vertices $\tilde{v} \in V$ have the same reachability $r(v)$ corresponding to Eq. (11.19). Since Eq. (11.21) corresponds to the definition of a strongly connected component, efficient algorithms from literature [11–14] might be applied such that the complexity to remove all cycles becomes linear.

For small problems, this reduction of complexity may not be significant, in particular because the timing analysis is done once at design time. However, an efficient fixed-point iteration approach becomes highly important if one or more of the following attributes hold:

- Analysis of large real-world examples with hundreds of components resulting in a high number of timing entities.
- Detailed modeling of also different (software) layers resulting in a high number of timing entities.
- Timing analysis is applied within a DSE resulting in a high number of independent timing analysis calculations.

In this case, the presented approach significantly outperforms known global fixed-point iteration approaches. An evidence of the benefits of the fine-grained fixed-point iteration is given in the experimental results in the following section.

11.4 Experimental Results

In order to give evidence of the efficiency of the proposed approach, a case study is presented. All following experiments were carried out on an Intel Core 2 Quad 2.66 GHz machine with 3 GB RAM.

11.4.1 Automotive Case Study

We consider an automotive network exploration case study. The network architecture consists of 15 ECUs, connected via 2 CAN buses, 1 FlexRay bus, and a central gateway. The 9 sensors, and 5 actuators are connected via LIN buses to the ECUs. An application consisting of four functions, an *adaptive cruise control* (ACC), a *brake-by-wire* (BW), an *air conditioning function* (C1), and a *multimedia control* (C2), with 46 processes and 42 messages in total, is mapped to the given architecture. The functions and their real-time end-to-end constraints are listed in Table 11.1.

The functions are distributed according to state-of-the-art real-world networks where the ACC is implemented in the FlexRay sub-network, BW and C1 are implemented in one of the CAN sub-networks, and C2 is implemented over both CAN

Table 11.1 Detailed information about the functions of the used case study in terms of numbers of processes and messages as well as the maximal latency of each function

Function	#Processes ($\|P\|$)	#Messages ($\|C\|$)	Max. latency [ms]
ACC	18	17	100
BW	8	7	50
C1	9	8	250
C2	10	9	150

sub-networks. This is regarded as the reference implementation. The reference implementation has the hardware cost of 216.80 € and an energy consumption of 11,745 mA and fulfills all real-time constraints. Additionally, several mapping and resource alternatives are added to enable an effective DSE.

11.4.2 Design Space Exploration Results

To illustrate the advantages of the DSE and the presented timing analysis, the automotive network is optimized in terms of the hardware cost in Euro (€) and energy consumption in milliamperes (mA). The hardware costs are approximated by a linear function based on the cost per resource whiles additional costs for wiring, etc. are neglected. The energy consumption is approximated by a non-linear energy model based on the average utilization of the ECUs. The timing constraints are not linearizable and have to be handled by the EA separately such that implementations that do not fulfill these constraints are discarded. The DSE includes a concurrent optimization of parameters such as the priorities of the processes and messages as well as the scheduling of the messages on the static or dynamic segment of the FlexRay bus.

The optimization requires 3,511 s using the fine-grained fixed point iteration. Within the optimization process, the EA obtains 5,075 implementations that require a timing analysis. Note that an exploration with the dependency-based approach requires 3 hours and 120 s, leading to a significant speed-up already for the presented small case study. The plain global fixed point iteration requires more than one day (after which it was aborted). The results of the optimization are illustrated in Fig. 11.6. Four non-dominated high quality implementations are found improving the reference implementation in both objectives, hardware cost and energy consumption. The found implementations decrease the hardware cost by 11.8–15 % while the energy consumption is decreased by about 3.6–8.7 % at the same time.

11.4.3 Timing Analysis Results

In the following, we want to focus on the reference implementation and just vary the priorities and schedules to illustrate the strongly varying runtimes of the timing analysis approaches even for a small case study. Here, 100 different configurations

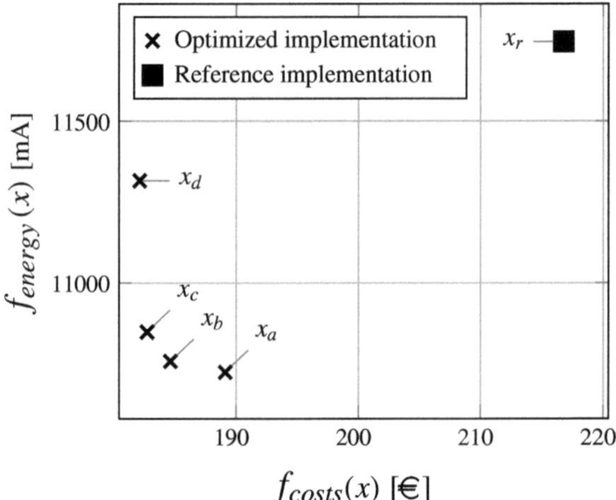

Fig. 11.6 The two dimensional plot of the optimization results and reference implementation of the automotive exploration case study

are evaluated. A comparison to the plain global fixed point iteration is omitted due to the long runtimes of this method. The runtime of the global dependency-based approach is 192 s. The fine-grained approach including the generation of the acyclic dependency graphs requires only 70 s and, thus, improves the runtime of a single evaluation by a factor of approximately 2.75 on average. The runtimes for all 100 evaluations are illustrated in Fig. 11.7. The plot shows that in many cases the runtime is approximately improved by a factor of two to four. On the other hand, the fine-grained fixed point iteration is also more than 8 times faster for some test cases even for this small case study.

The presented case study is rather small compared to real-world systems. For instance, in the automotive area, state-of-the-art architectures consist of up to 100 ECUs connected via several buses with hundreds of tasks and messages. Using the fine-grained fixed point iteration for such large systems shall improve the runtime of the timing analysis even more significantly. Moreover, the growing amount of computationally expensive timing analysis for some components that are based on Integer Linear Programming (ILP) [15] or model checking [16] require an efficient fixed point iteration approach in case of cyclic dependencies.

11.5 Concluding Remarks

This chapter presents an efficient Design Space Exploration (DSE) using a fast timing analysis method. For the timing analysis, a timing entity graph is constructed and partitioned to achieve a fine-grained fixed point analysis. In future work, the proposed approach shall be combined with more complex timing analysis approaches that rely

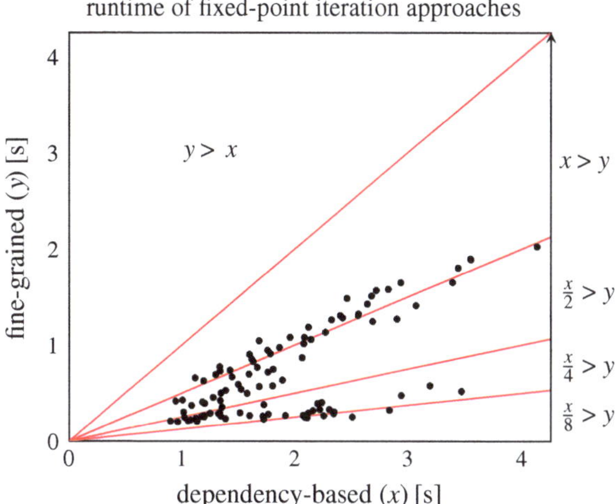

Fig. 11.7 Runtime comparison of different fixed point iteration approaches for 100 different priority-configurations for the reference implementation. Each dot specifies the runtime of the respective methods to determine the timing properties for a given implementation

on ILP approaches or model checking. In this case, a fast timing analysis becomes inevitable if cyclic dependencies exist to significantly minimize the runtime of a DSE.

Acknowledgments This work was financially supported in part by the Singapore National Research Foundation under its Campus for Research Excellence And Technological Enterprise (CREATE) programme.

References

1. Richter, K., Ziegenbein, D., Jersak, M., Ernst, R.: Model composition for scheduling analysis in platform design. In: Proceedings of the 39th Conference on Design Automation (DAC 2002), pp. 287–292 (2002)
2. Anssi, S., Albers, K., Dörfel, M., Gérard, S.: ChronVAL/ChronSIM: a tool suite for timing analysis of automotive applications. In: Proceedings of the Conference on Embedded Real-time Software and Systems (ERTS 2012) (2012)
3. Chakraborty, S., Kunzli, S., Thiele, L.: A general framework for analysing system properties in platform-based embedded system designs. In: Proceedings of the Conference on Design, Automation and Test in Europe (DATE 2003), pp. 190–195 (2003)
4. Blickle, T., Teich, J., Thiele, L.: System-level synthesis using evolutionary algorithms. Des Autom. Embed. Syst. **3**(1), 23–62 (1998)
5. Lukasiewycz, M., Streubühr, M., Glaß, M., Haubelt, C., Teich, J.: Combined system synthesis and communication architecture exploration for MPSoCs. In: Proceedings of the Conference on Design, Automation and Test in Europe (DATE 2009), pp. 472–477 (2009)
6. Lukasiewycz, M., Glaß, M., Haubelt, C., Teich, J.: SAT-decoding in evolutionary algorithms for discrete constrained optimization problems. In: Proceedings of CEC '07, pp. 935–942 (2007)

7. Tindell, K., Burns, A., Wellings, A.: Calculating controller area network (CAN) message response times. Control Eng. Pract. **3**, 1163–1169 (1995)
8. Künzli, S., Hamann, A., Ernst, R., Thiele, L.: Combined approach to system level performance analysis of embedded systems. In: Proceedings of the 5th IEEE/ACM International Conference on Hardware/Software Codesign and System, Synthesis (CODES+ISSS 2007), pp. 63–68 (2007)
9. Schioler, H., Jessen, J., Nielsen, J.D., Larsen, K.G.: Network calculus for real time analysis of embedded systems with cyclic task dependencies. In: Proceedings of the 20th International Conference on Computers and Their Applications (CATA 2005), pp. 326–332 (2005)
10. Jonsson, B., Perathoner, S., Thiele, L., Yi, W.: Cyclic dependencies in modular performance analysis. In: Proceedings of the 8th ACM International Conference on Embedded software (EMSOFT 2008), pp. 179–188 (2008)
11. Aho, A.V., Hopcroft, J.E.: Ullman. Data Structures and Algorithms. Addison-Wesley, J.D. (1983)
12. Cheriyan, J., Mehlhorn, K.: Algorithms for dense graphs and networks on the random access computer. Algorithmica **15**(6), 521–549 (1996)
13. Sedgewick, R.: Algorithms in C, Part 5: Graph Algorithms. Addison-Wesley (2002)
14. Tarjan, R.: Depth-first search and linear graph algorithms. SIAM J. Comput. **1**(2), 146–160 (1972)
15. Pop, T., Pop, P., Eles, P., Peng, Z., Andrei, A.: Timing analysis of the FlexRay communication protocol. Real-Time Syst. **39**(1), 205–235 (2008)
16. Lampka, K., Perathoner, S., Thiele, L.: Analytic real-time analysis and timed automata: a hybrid method for analyzing embedded real-time systems. In: Proceedings of the 9th ACM International Conference on Embedded software (EMSOFT 2009), pp. 107–116 (2009)

Chapter 12
Design and Evaluation of Future Ethernet AVB-Based ECU Networks

Michael Glaß, Sebastian Graf, Felix Reimann and Jürgen Teich

Abstract Due to ever-increasing bandwidth requirements of modern automotive applications, Ethernet AVB is becoming a standard high-speed bus in automotive E/E architectures. Since Ethernet AVB is tailored to audio and video entertainment, existing analysis approaches neglect the specific requirements and features of heterogeneous E/E architectures and their applications. This chapter presents a virtual prototyping approach to consider Ethernet AVB in complex E/E architectures, reflecting key features such as static routing and stream reservation, fixed topology, and real-time applications. A comparison with a timing analysis on case studies from the automotive domain gives evidence that the proposed simulation technique delivers valuable bounds for complete sensor-to-actuator chains, enabling automatic system synthesis and design space exploration approaches.

12.1 Future Communication Media for ECU Networks

Modern automotive E/E architectures consist of more than 100 *Electronic Control Units* (ECUs), numerous sensors and actuators, as well as a complex communication architecture comprising a heterogeneous system of (field) buses that are connected via gateways. Already today and even more in the near future, the bandwidth of well-

M. Glaß (✉) · S. Graf · F. Reimann · J. Teich
Hardware/Software Co-Design, Friedrich-Alexander-Universität Erlangen-Nürnberg, Erlangen, Germany
e-mail: glass@cs.fau.de

S. Graf
e-mail: sebastian.graf@cs.fau.de

F. Reimann
e-mail: felix.reimann@cs.fau.de

J. Teich
e-mail: teich@cs.fau.de

A. Sangiovanni-Vincentelli et al. (eds.), *Embedded Systems Development*, Embedded Systems 20, DOI: 10.1007/978-1-4614-3879-3_12,

established field bus technologies like LIN, CAN, or FlexRay will be exhausted by infotainment and advanced driver assistance functions. To cope with this problem, Ethernet as a well-established bus technology from the consumer electronics domain has gained attention due to its high bandwidth and low cost components, cf. [1]. In particular, focus is currently put on Ethernet AVB [2], a light-weight Ethernet extension towards enhanced Quality-of-Service (QoS) tailored to audio and video transmission, suiting the infotainment domain well, see [3, 4].

However, automotive E/E architectures are known for their heterogeneous applications with respect to both their criticality in the sense of real-time and safety requirements as well as their communication characteristics, see [5]. In particular, communication characteristics that range from streaming applications with high throughput demands share one or several communication media with real-time critical control applications that require low latencies and jitter for periodic messages. Given there exist suitable bus techniques for both domains, like MOST150 for streaming media or FlexRay for real-time applications, using these techniques for either domain exclusively results in highly increased cost and typically also a violation of other constraints such as the available mounting space. To avoid such a heterogeneity, Ethernet AVB with its prioritization and traffic shaping techniques known from the QoS domain is an option to unify several bus systems and integrate applications of mixed criticality. This requires sophisticated prototyping techniques to not only consider applications that fit the QoS concept within Ethernet AVB, but to also properly reflect the resulting delays and jitters for real-time applications.

Using Ethernet AVB as a case study, the work at hand investigates the use of a high-bandwidth bus with QoS characteristics within a heterogeneous networked embedded system. The carried out prototyping approach aims at analyzing the timing properties of the applications, given Ethernet AVB is employed as one of several bus systems in a heterogeneous E/E architecture. This work incorporates *Virtual Processing Components* (VPC) [6] and some extensions proposed in [7] as the basic prototyping technique. Furthermore, the approach aims at being compatible to existing VPC modeling and timing analysis approaches which are used in the case study to model other components for the sake of design space exploration of extensive E/E architectures, see for example [8]. Moreover, it is prospected that Ethernet AVB will be used without the optional dynamic stream reservation, but streams are statically configured as typical for the automotive domain. Thus, this work makes use of this fact and allows static as well as dynamic stream reservation.

The remainder of the chapter is outlined as follows: In Sect. 12.2, the related work for timing simulation and analysis of Ethernet AVB is discussed. The traffic shaping of Ethernet AVB is given in Sect. 12.3. Section 12.4 presents the basic model and the proposed virtual prototyping extensions for Ethernet AVB. Experimental results for a real-world case study are discussed in Sect. 12.5 while Sect. 12.6 concludes the chapter.

12.2 Related Work

Simulation-based approaches for standard Ethernet with focus on automotive application and communication-based end-to-end delays for Ethernet AVB are presented in [9–12]. All these approaches are based on the network simulator OMNet++ and allow a very accurate simulation of the networking traffic and the involved protocols. Moreover, they allow to simulate heterogeneous networking structures by combining different network models. Other simulation based approaches like [13] are able to analyze an AVB network very accurately, while focusing on frames on network level and, thus, neglect the impacts to the functionality. In contrast, the work at hand focuses on the functional behavior of the automotive system and the effects of the heterogeneous, underlying network on the involved applications. Especially the distributed implementation of complex functions and real-time requirements within the automotive domain show the importance of analyzing the systems timing and functionality at the system level,regarding the application in combination with the underlying architecture and communication infrastructure. This requires on the one hand an efficient simulation of the networking systems as well as the integration of timing chains and functionality.

Beside the simulation based approaches, real-time analysis approaches tackling the problem of guaranteed delays are increasingly important. Real-time analysis of standard Ethernet by employing and extending the commercially available tool SymTA/S [14] is proposed in [15, 16]. For Ethernet AVB, the IEEE standard [2] provides a formal worst case analysis for the local delay a frame may experience by higher priority *Credit-Based Shaper* (CBS) queues at a single queue. The analysis proposed in [17] considers Ethernet AVB streams over multiple communication hops. In [18], a network calculus model to compute the worst case of Ethernet AVB communication from sender to receiver is proposed. But, as distinct to our work, these approaches neglect the influences of the communication behavior to the functional behavior, especially the end-to-end delay. The work at hand is capable of analyzing complete sensor-to-actuator task chains by considering data at the level of frames as well as messages, but only delivers average-case and typical values, so that it is common to use the presented simulation technique in combination with a real-time analysis. Moreover our approach is also capable to gather the distribution of the delays occurring at network layer as well as at application layer.

Furthermore, the proposed virtual prototyping approach is capable of investigating the aspect of *overreserved bandwidth* as a means to satisfy real-time requirements of control applications in QoS-driven network techniques.

12.3 Fundamentals

Ethernet AVB provides a QoS enhancement for Ethernet by (a) a synchronization protocol, (b) a stream reservation protocol for dynamic message streams, and (c) the forwarding and queuing enhancements for the hardware. Given (b) is not required

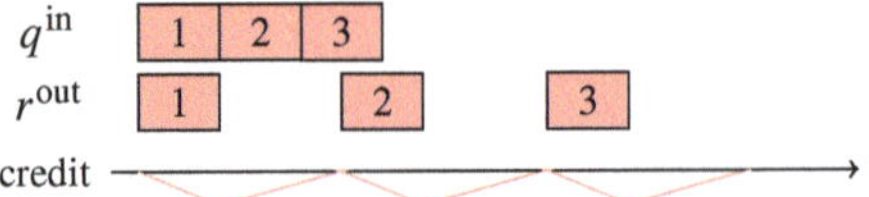

Fig. 12.1 A simple example of a Credit-based Shaper (CBS). Three frames arrive in a burst. If the credit is ≥ 0, one frame is sent while negative credit is accumulated as long as the frame is transmitted. Once the transmission ends, the credit increases again until it reaches 0, enabling to send the next frame

in automotive E/E architectures, this work focuses on (c). AVB enhances the VLAN standard [19] which introduces traffic classes to Ethernet: Each network port has up to eight output *queues* which can be addressed by a priority field in the Ethernet frame header, the *VLAN tag*. Moreover, AVB introduces an optional *Credit-based Shaper* (CBS) to an output queue, aiming at reducing bursts. If a CBS queue transmits a frame, the credit is reduced linearly depending on the size of the frame and the reserved bandwidth of the queue, see Fig. 12.1. A second frame can only be sent if the credit has regenerated again, i. e., is greater or equal to 0. If the transmission of a frame is delayed, either by a lower priority frame which is currently under transmission or by higher priority CBS queues, the credit aggregates (becomes positive) and, thus, more than one frame may be transmitted at once as long as the credit is positive. Thus, bursts are still present in AVB networks. If no frame is enqueued, remaining positive credit is discarded.

But, as multiple frames and multiple queues compete for the same network port, frames might experience jitter. The jitter j_{Queue} of a message m on an AVB queue q is defined as the maximum variation from the best case delay $d_{\text{Queue}}(m, q)$. It consists out of three different types: (1) the *head-of-line blocking* j_{hol} introduced by a frame of a lower traffic class, which is currently under transmission, (2) the *high-priority jitter* j_{hp} introduced by messages of higher priority transmitted in lower traffic classes, and (3) the *fan-in jitter* $j_{\text{fan-in}}$ introduced by messages of the same traffic class which are also in the output queue. Thus, the jitter of a message is given as follows:

$$j_{\text{Queue}}(m, q) = j_{\text{hol}}(m, q) + j_{\text{hp}}(m, q) + j_{\text{fan-in}}(m, q) \tag{12.1}$$

The transmission of a single Ethernet frame is not preemptive. Thus, a frame may be delayed by a frame currently under transmission at a lower priority queue. Here, j_{hol} is the time a maximum-sized frame of higher traffic classes (lower priority) requires for transmission, see Fig. 12.2. Fan-in jitter $j_{\text{fan-in}}$, see Fig. 12.3, is the most important jitter introduced by AVB networks: As messages of the same traffic class may be put in the output queue in arbitrary order, all frames already enqueued need to be transmitted according to the reserved bandwidth before the message under examination is transmitted. Thus, the time this amount of data requires for transmission with the reserved bandwidth of the traffic class is the desired fan-in jitter.

Our simulation based approach abstracts from the low level effects and is able to provide the overall distribution of delays that could occur in the system by analyzing

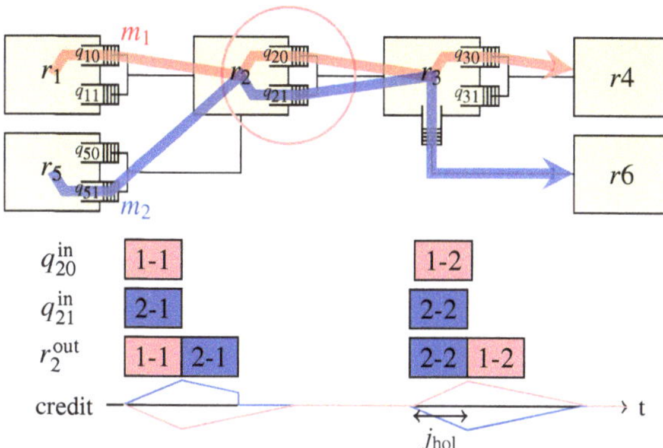

Fig. 12.2 The message m_1 may be delayed by j_{hol} by the low priority message m_2 at r_2. Here, the first frame of m_1 arrives concurrently with the frame of m_2 and is transmitted immediately. Thus, its credit falls below 0 while the credit of delayed queue q_{21} increases. The second frame of m_1 arrives slightly after that of m_2 and, thus, has to wait till the transmission is finished. In the meantime, the credit of q_{21} increases due to the waiting frame

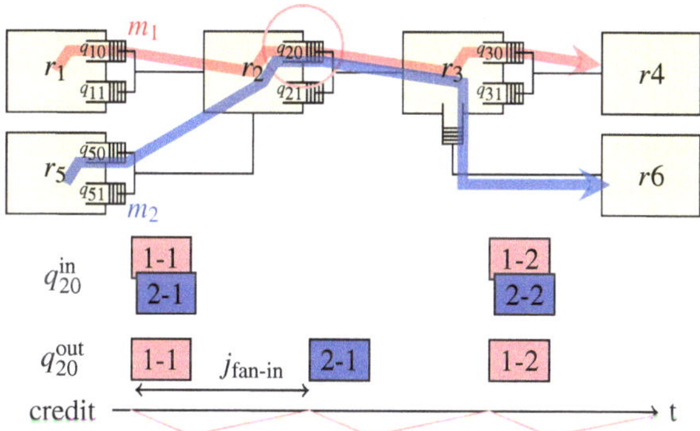

Fig. 12.3 The messages m_1 and m_2 share the same traffic classes and, thus, may be delayed by j_{fan-in} depending on the order they arrive at the output queue. Instead of being sent immediately after m_1, m_2 has to wait until the credit has regenerated

the timing properties of a whole message possibly consisting of multiple frames at application level. As can be seen in the case study, the delivered variations due to introduced clock drift and local task jitters are able to project the real (simulated) behavior to analytical best- and worst case timings delivering tight timing values.

The simulation based virtual prototyping approach used in this work is based on the principles of the Y-chart approach introduced by BALARIN et al. [20], which clearly separates the functionality from the architecture and fits the automotive domain well. In our case, the functionality is modeled as an executable specification in SysteMoC [7] that merges the abilities of SystemC with well defined models of computation. It allows an *actor*-based modeling of the functionality with dedicated communication through *channels*.

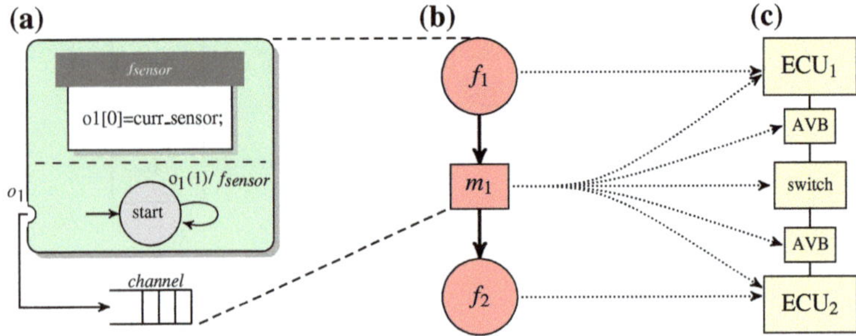

Fig. 12.4 The overall modeling approach consists of three parts: **a** Functionality is modeled by multiple SysteMoC actors communicating via channels. **b** The actors and channels are transformed to the functional network, consisting of tasks (actors) and messages (channels). To create the virtual prototype, the functional network is mapped to **c** the component network that models sensors/actuators, ECUs, as well as communication infrastructure

An actor, see Figure 12.4 a, can consist of *actions* implementing its various functional parts. Additional to that, the activation and communication behavior of the actor is controlled by a finite state machine. Multiple actors are connected via channels to a functional network, see Figure 12.4 b. The functional network is mapped to *Virtual Processing Components* (VPC), which model the temporal behavior caused by arbitration and scheduling technique of the represented hardware component as the architectural part of the Y-Chart approach, see Figure 12.4 c. Both parts are modeled independently and, thus, they have to be coupled to allow a combined simulation. In case of our work, the whole coupling between various system threads is implemented with the help of events (see Fig. 12.5).

During the simulation of the functional model, after each execution of an *action* (state transition 1 in Fig. 12.5), the framework uses the coupling to the VPC framework to inform the resource about the new task and to start the timing simulation, e. g., determine the end of the current execution. The component itself is informed about the new *task*[1] that has to be executed on the component and, thus, tries to determine the point of time, the execution is finished. For that, the component knows the current *execution delay* the tasks would have, if no preemption, resource contention, and QoS happens or is applied. Moreover, to determine the current execution state and the timing behavior, each component has a fixed *scheduling strategy* modeling its arbitration and resource usage mechanism (e. g., the software scheduler within an operating system or the arbitration protocol of a bus system).

This scheduler is called by the component each time a new task arrives (see state transitions 2 and 5 in Fig. 12.5), a task has consumed its delay (see state transitions 6 and 9 in Fig. 12.5), or in case of other timing events, e. g., due to time-triggered scheduling occur. In case a task gets exclusive access to the computational or communication resource (see state transition 3 in Fig. 12.5), the scheduler returns an

[1] *Task* in this case could either be a functional task executed on or a message that is currently transmitted over the component.

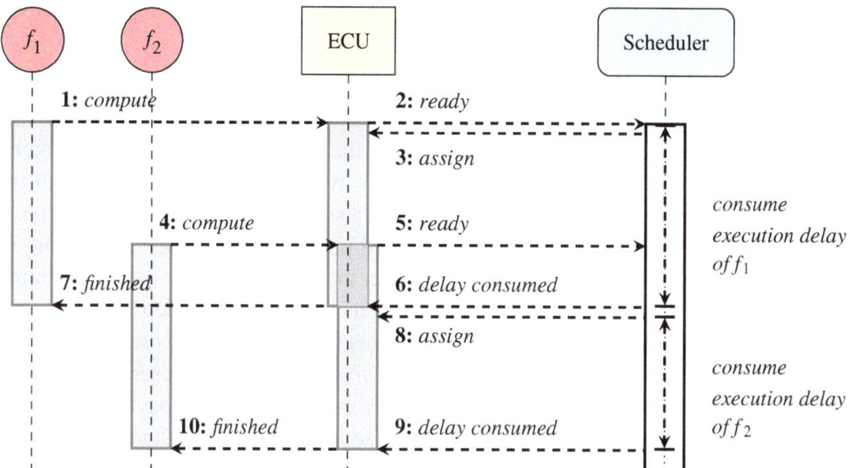

Fig. 12.5 The coupling between the functional and the component network. An actor's method (*action*) is atomically executed at time of activation (e. g., f_1, f_2). Then, the corresponding VPC starts the timing simulation (1, 4) for this task and returns when the task has been finished and its given execution delay has been consumed (7, 10). Moreover, the related scheduling mechanism (in the example shown: *first come first serve*) is informed about a new task (2, 5), assigns the task in case the resource is available (3, 8), and determines the end of the execution delay (6, 9). In case of other scheduling mechanisms like *priority based preemption*, the timing behavior might be different

assign. After consuming the required delay, the VPC informs the framework that the action was executed successfully and, thus, the simulation can advance, e. g., inform the functional network that the action has performed and finished its execution on the component or advance the static route of a message (see state transitions 7 and 10 in Fig. 12.5). Due to the separation of functionality and architecture, multiple tasks might be executed on the same component in parallel. Thus, it is very important that the scheduling mechanism is modeled as fine-grained as necessary, but, to lower the simulation overhead, as abstract as possible.

To extend the existing framework to support QoS mechanisms, this work introduces a VPC model for Ethernet AVB and employs existing VPC-like E/E architecture modeling extensions (including models for ECUs, sensors, actuators, and other buses) from [7]. Moreover, techniques like stream reservation and overreservation are integrated. Additional to this, dynamic stream reservation as presented in [21] could also be used.

12.4 VPC Model

As already mentioned, architectural components, especially their scheduling and arbitration behavior, are modeled with the help of parameterized VPCs. Moreover, as today's automotive systems consist of many heterogeneous ECUs and various networking components, the complete E/E architecture must be modeled with multiple instances of VPCs. Furthermore, special parameters like the scheduling used or static stream reservation and configuration have to be annotated.

Moreover, some restrictions are taken into account: In the following, we distinct between *messages* $m \in M$, which model the complete amount of data being sent between tasks in an application, and *frames*, which include the overhead of the UDP/IP and Ethernet protocol headers and have a size up to the *Maximum Transmission Unit* (MTU), which is defined by the used physical layer. This distinction, of course, enables to define messages of sizes larger than enabled by the communication infrastructure. The derived virtual prototyping technique is capable of considering this fact to achieve a reasonable analysis of the overall system incorporating *message fragmentation* in case of production of multiple *tokens* (each is representing a frame) that are transmitted over the *channel*. For each message its size (payload) p_m and the traffic class c_m is given. Let p_m be the size of message m. Then, $s(m)$ is the effective size of message m with all headers and footers of the used network stacks. For example, for UDP/IP over Ethernet, $s(m)$ for a message with a given payload p_m, is given by (see [15]):

$$s(m) = 58 \cdot \left\lceil \frac{p_m + 8}{1472} \right\rceil + p_m + 8 + \max\left(0,\ 26 - ((p_m + 8) \bmod 1472)\right) \quad (12.2)$$

Moreover, $s_f(m)$ is the size of frame f in case message m is larger than the MTU and, thus, needs to be fragmented. The best case delay $d_{\text{Queue}}(f)$ of frame f is determined by the transmission delay as follows:

$$d_{\text{Queue}}(f) = \frac{1}{\text{linespeed}} * s_f(m) \quad (12.3)$$

A traffic class $c : M \mapsto C; C = \{0, \ldots, 7\}$ comprises distinct messages and provides specific QoS mechanisms, see [19]. A network resource may provide up to eight different outbound queues, where a message m with $c(m) = 0$ is best effort traffic (lowest priority). The task of the network designer is to map each message to one of the eight traffic classes. In particular, IEEE 802.1Qav [2] provides means to map each class $c(m)$ to the currently available queues.

In the following, we describe how the integration of Ethernet AVB with its credit-based traffic shaping and the best effort queues can be efficiently integrated as a new component to the Virtual Processing Component framework. Thus, we (1) describe the general separation, (2) transform the CBS behavior to a scheduling behavior usable within an event-based simulation, and (3) integrate best-effort behavior for non reserved streams.

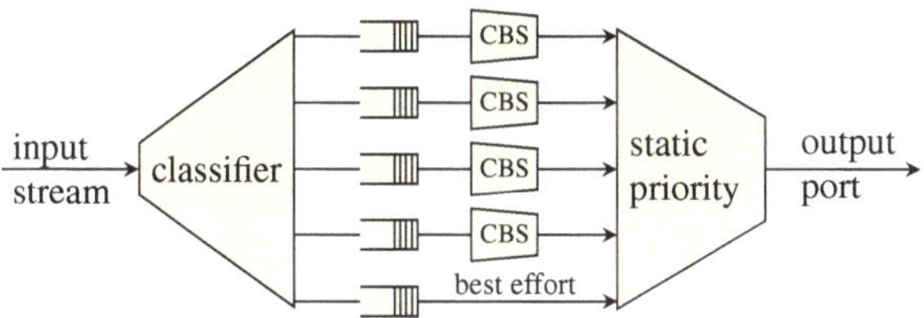

Fig. 12.6 Model of an AVB output port with four CBS queues and one queue for best effort traffic

12.4.1 AVB Scheduling

As described before, the scheduling approach of Ethernet AVB uses two different kinds of output queues: multiple queues with *credit-based shaping* (CBS) for reserved communication and one *best effort* (BE) queue for ordinary messages. Therefore, the standard provides a regulation, how to combine these different kinds of strategies as a scheduling policy. In case of AVB, a fixed order by means of the VLAN priority tag was defined. A higher priority instance (typically a CBS-queue) is allowed to access the link if (a) it has at least a waiting packet ($framecount > 0$), (b) the instance itself is allowed to send a packet due to its internal QoS-settings (e. g. $Cred_c \geq 0$), and (c) no higher priority queues fulfill the requirements (a,b).

In case of our modeling, the scheduler itself integrates an ordered list of all instances, which are requested in a top down order as the primary scheduling policy. Furthermore, as depicted in Fig. 12.6, the scheduler combines the classification of incoming frames as well as the handling of the various queues and the arbitration of the output port.

12.4.1.1 AVB CBS Scheduling Approach

Ethernet AVB ensures a quality of service by a CBS algorithm, which is partly defined in [17]: The *idleSlope* iS_c denotes the reserved bandwidth in the percentage of the line speed for the traffic class c. Thus, iS_c is defined as the sum of the bandwidth fractions, given as $bw(m)$ for each message $m \in M$ mapped to c:

$$iS_c = \sum_{\forall m \in M,\ c(m)=c} bw(m) \tag{12.4}$$

If overreservation is considered, e. g., to reduce the jitter, iS_c can be increased if the overall reserved bandwidth is still less than the linespeed.

The credit counter $Cred_c$ increases with the *idleSlope* while a frame is ready for transmission, but is not allowed to access the outgoing port (see Sect. 12.4.1) due to insufficient credit or usage of the port by another traffic class. Moreover, if the traffic class c sends a frame, its credit is decreased and must be regenerated in order to re-access the link (in case it became negative). Thus, $Cred_c$ can be calculated as following, given that Δt is the time interval between the current simulation time and the last scheduling decision at this component:

$$Cred_c = \begin{cases} Cred_c + iS_c * \Delta t & \text{if } framecount > 0 \; and \; link \; is \; used \\ Cred_c - (1 - iS_c * \Delta t) & \text{else if } Class \; c \; uses \; the \; link \\ Cred_c + iS_c * \Delta t & \text{else if } Cred_c < 0 \; and \; (Cred_c + iS_c * \Delta t) < 0 \\ 0 & \text{else if } (Cred_c + iS_c * \Delta t) \geq 0 \\ 0 & \text{else if } Cred_c > 0 \; and \; framecount == 0 \\ Cred_c, & \text{else.} \end{cases} \quad (12.5)$$

In case of real hardware and also in case of other simulation approaches, this requires an ongoing update of the available credit counters for all queues, as each transmitted Byte causes multiple changes to the various credit counters, as could also be seen in Fig. 12.1. In case of multiple CBS queues and large packets, this may result in heavy additional control usage of the resource. But, as Ethernet-based networks do not support preemption, i. e., a frame is always transmitted completely, thus, this kind of behavior is not necessary.

In case of our event based simulation, this handicap can be omitted due to the clear separation between functionality, communication, and architecture. As the scheduler is only called in case of changes to the set of running and waiting tasks or messages (e. g. arriving of a new task or finishing of an existing task), it is sufficient to update the credit counters at these points in time. Furthermore, only in case the resource gets unused, new scheduling decisions must be made, requiring a new examination of the available credits. This substantial lowers the computational overhead of the scheduling during the simulation and, thus, speeds up the simulation and evaluation process using the virtual prototype.

Moreover, as (nearly) each of the eight traffic classes might be implemented as a CBS queue, multiple instances of this approach must be able to be instantiated and, of cause, be updated based on the related bandwidth allocation. Thus, a mapping of each reserved stream to one of the given traffic classes by a classifier is mandatory (see also Sect. 12.4).

12.4.1.2 Best Effort Scheduling Approach

Beside credit-based shaping and the use of guaranteed bandwidth, Ethernet AVB also supports legacy frame transmission. As common in typical automotive networks, only a single instance of the best effort queue is available. The local scheduling strategy of the BE queue follows the priority-based scheduling by using the VLAN priority field as input. If only the BE queue is used, the network as well as the simulation decays to standard Ethernet.

In case of integration into our VPC model, the queue can be implemented as a priority queue providing all the necessary abilities for modeling the BE behavior.

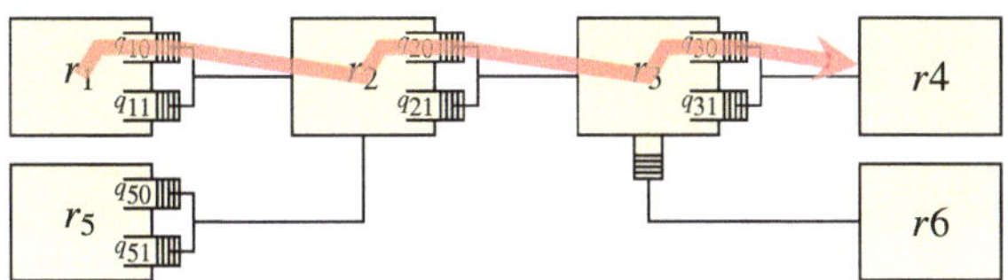

Fig. 12.7 A message m is sent by a task running on resource r_1. It is transmitted to the output queue q_{10} of the resource r_1. From there, it is routed over switch r_2 and its output queue q_{20} via switch r_3 and its output queue q_{30} to the resource r_4 on which the receiving task is executed

Each time, no CBS queue is able to send a frame, the top element of the queue is transmitted via the corresponding link. Again, due to performance issues, this is only required in case the link gets unused.

Moreover, in case of reducing the diversity of the overall scheduling methodology, this approach can also be transformed to be implemented with the help of the CBS queue approach presented before. It can be shown that the BE behavior can be implemented with the CBS approach by allocating the line speed for the lowest traffic class, i. e., the credit does never increase and is also never decreased. To avoid an overflow of the credit,[2] it is reset after it exceeds a given threshold. Thus, we use another instance of the CBS queue model and set the reserved bandwidth $idleSlope_c$ for channel c_{BE} to 1.

12.4.2 Overall Ethernet AVB Model

As the presented AVB abstraction (Sect. 12.4.1) models the timing behavior of an AVB port, we also need models for active networking components like an Ethernet switch. But, as a switch is something like a collection of AVB ports which are connected by a highspeed backplane bus, most of the network can again be build upon the basic VPCs as presented before. One example for a possible system configuration can be seen in Fig. 12.7, where each link between two resources is implemented as a dedicated VPC. Note that the handling of multiple queues is also integrated in the scheduler and, thus, the example consists of four computational resources (r_1, r_4, r_5, r_6), two switch instances (r_2, r_3) and five AVB port instances (all links between resources as output ports). The VPC of an ECU like r_1 depends on the scheduler used there. Typically, priority-based or first-come-first-serve scheduling is applied. Input queues of communication controllers can be ignored, as they do not incorporate any additional delay or jitter not respected in switches or output queues, cf. [2]. A possible result of the message shaping based on the architecture shown in Fig. 12.7 can be seen in Fig. 12.8.

[2] A credit overflow may happen in case of a permanent overloaded link by upper class CBS queues, but should not be allowed, because it is a faulty system design.

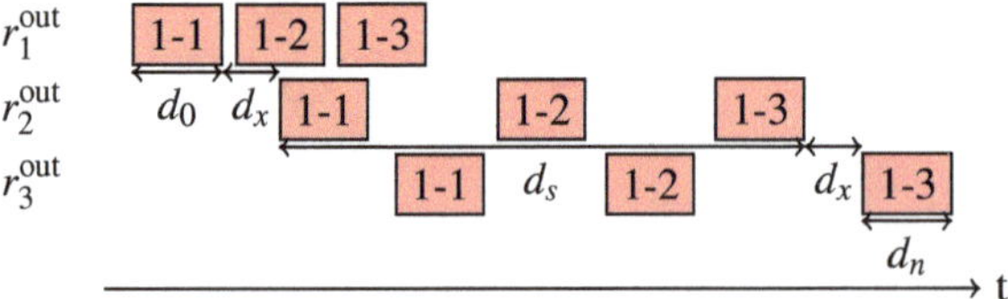

Fig. 12.8 The message m of Fig. 12.7: Let queue $q_{20} = q_s$ have the lowest reserved bandwidth while q_{10} and q_{30} have a higher reserved bandwidth. d_x is the delay incorporated by the switch fabrics

The behavior of the involved network stacks at the sending and receiving ECUs is not part of the AVB standard and, thus, has to be adapted to the specific operating system of the target component. However, as measures of typical hard- and software in [22] show, most network layers at the senders and receivers, namely sockets, IP, and MAC layer introduce only a static delay to the transmission or a delay which depends on the size of the frame (e. g., the UDP layer). This can be seamlessly modeled as an additional static delay by standard VPC components introducing configured *delays* to the used static route, e. g. they are added as a static offset in the calculation of the transmission delay at the first and last hop.

The modeling of the backplane of the switch is discussed in the following. The scheduling policy of switches is not standardized, neither for Ethernet AVB nor for standard Ethernet, see [23, 2]. Thus, the response time analysis and simulation modeling for switches depends on the used hardware. However, the typical switches in embedded systems use a backplane bus which is non-waiting, i. e., the bandwidth of the backplane bus is higher than the sum of all outbound queues. To model our hardware, we use an existing model for round robin scheduling and a user-definable backplane bandwidth and, thus, the switch is modeled as a usual VPC instance.

12.5 Case Study

To show the applicability and accuracy of the proposed analysis method a real-world automotive subsystem with Ethernet AVB for sensor-to-actuator task chains in a heterogeneous E/E architecture is investigated. The case study is modeled with the CASE tool PREEvision [24] for designing E/E architectures and semi-automatically transformed to our simulation models and a real-time analysis approach.

The case study models a typical automotive subnetwork consisting of 48 tasks running on 12 different ECUs, 43 messages routed over 3 switches, and 12 LIN buses used for the low cost connection of sensors and actuators. Table 12.1 gives additional information about the applications. Note that the messages of app_3 and app_5 are larger than the MTU and, thus, will be fragmented. The message routings, traffic classes, and the *idleSlopes* for the CBS queues are determined by employing an automatic

Table 12.1 Applications of the case study

	#tasks	#messages	payload [Byte]	period [ms]	deadline [ms]
app_1	1	0	–	4	–
app_2	1	0	–	3	–
app_3	2	1	1540	4–5	–
app_4	17	18	4–480	1–10	10
app_5	4	3	184320	5–40	100
app_6	10	9	4–96	1–5	10
app_7	13	12	20–1200	1–5	5

Table 12.2 Simulation runtimes [s]

# frames	# hops				
	3	5	11	21	41
10^3	0.07	0.09	0.16	0.28	0.50
10^4	0.64	0.84	1.42	2.44	4.94
10^5	6.25	8.26	13,94	24.40	49.10
10^6	63.61	81.86	143.8	243.5	489.5

design space exploration framework [25, 26]. To evaluate the timing behavior, the simulation is configured automatically and simulates a given scenario of 10 seconds. The execution of the simulation for this case study lasts about 14 seconds on a 3 GHz single-core processor. For comparison, Table 12.2 provides runtime measurements of a synthetic scenario with varying numbers of hops and frames. It shows that the runtime scales linearly in the number of frames and used hops.

Figure 12.9 shows the end-to-end latency for app_5, which is a video processing application of an advanced driver assistance system, where two video streams from two cameras are merged at a central *Electronic Control Unit* (ECU) and sent to the display. The histogram shows the latency for the stream starting at one of the cameras till the image is displayed. These three video streams result in 378 distinct Ethernet frames per video frame. In contrast to related work, the proposed modeling needs only to consider three messages per network resource but, however, is nonetheless able to deliver a good distribution between tight bounds for the best and worst case calculated by the real-time analysis. Moreover, the low jitter is remarkable and not achievable with standard Ethernet although using 100 MBit/s links.

Figure 12.10 shows the control application app_4, which messages are also routed over the two switches used by app_5. The model considers not only Ethernet AVB here, but the complete sensor-to-actuator chain including scheduling on the ECUs, barrier synchronization in the application (sensor fusion), etc. It highlights the applicability of the proposed overall virtual prototyping approach that is now capable of considering the effects of a QoS high-speed bus within a heterogeneous E/E architecture. Additionally, this enables automatic optimization approaches [25], particularly a

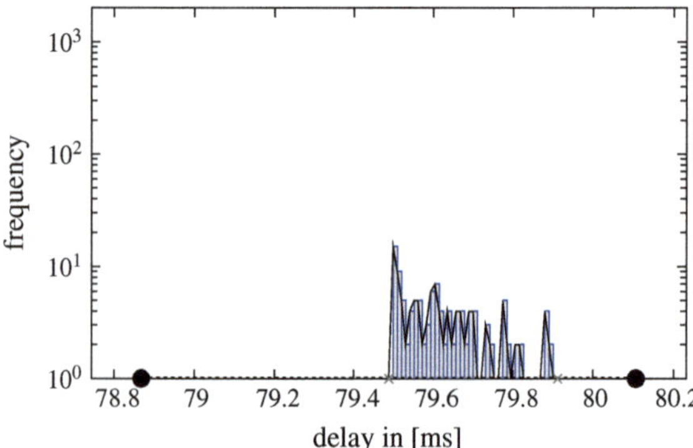

Fig. 12.9 The histogram shows the results of the timing simulation of the video processing application app_5. The black dots mark the best and worst case, respectively, as given by the real time analysis

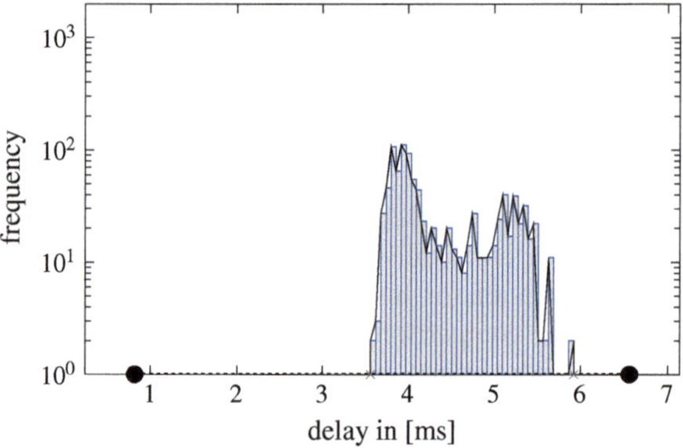

Fig. 12.10 The histogram shows the results of the timing simulation of the control application app_4 which messages are partly routed over the same network components as app_5. The black dots mark the best and worst case, respectively, as given by the real time analysis

parameter optimization. Moreover, it allows to evaluate the quality of results provided by a real-time analysis approach.

12.6 Conclusion

Ethernet AVB is a light-weight Ethernet extension to enhance its QoS aspects. In this work, a virtual prototyping approach based on virtual processing components which reflects features of future AVB-based E/E architectures such as static routing and stream reservation, fixed topology, and real-time applications is presented. The work at hand gives evidence that (a) the proposed methodology enables an efficient modeling of Ethernet AVB, especially the CBS queues and delivers timing deviations that fit well to tight bounds calculated by a real-time analysis and allows an analysis of the systems' functionality and (b) that Ethernet AVB is superior to standard Ethernet if designed well. A case study from the automotive domain gives further evidence of the applicability of the proposed approach to the analysis of system-wide complete sensor-to-actuator chains. The proposed technique is, thus, an important step towards automatic optimized E/E architecture design in the presence of high-speed bus systems with enhanced QoS capabilities.

Acknowledgments This work is supported in part by the German Federal Ministry of Education and Research (BMBF) under project 01BV0914-SEIS.

References

1. Glaß, M., Herrscher, D., Meier, H., Piastowski, M., Schoo, P.: SEIS - Security in Embedded IP-based Systems. ATZelektronik worldwide **1**, 36–40 (2010)
2. The Institute of Electrical and Electronics Engineers, Inc.: IEEE Standard for Local and metropolitan area networks - Virtual Bridged Local Area Networks, IEEE Std 802.1qav-2009 edn. (2009)
3. Lim, H., Volker, L., Herrscher, D.: Challenges in a future IP/Ethernet-based in-car network for real-time applications. In: Design Automation Conference (DAC), pp. 7–12 (2011)
4. Streichert, T., Buntz, S., Leier, H., Schmerler, S.: Short and long term perspective for Ethernet for vehicle-internal communication. 1st Ethernet & IP Automotive Techday (2011)
5. Sangiovanni-Vincentelli, A., Di Natale, M.: Embedded system design for automotive applications. IEEE Computer **40**(10), 42–51 (2007)
6. Streubühr, M., Gladigau, J., Haubelt, C., Teich, J.: Efficient Approximately-Timed Performance Modeling for Architectural Exploration of MPSoCs. In: Advances in Design Methods from Modeling Languages for Embedded Systems and SoC's, vol. 63, pp. 59–72. Springer (2010)
7. Graf, S., Streubühr, M., Glaß, M., Teich, J.: Analyzing automotive networks using virtual prototypes. In: Proceedings of the Automotive meets Electronics (AmE), pp. 10–15 (2011)
8. Lukasiewycz, M., Glaß, M., Haubelt, C., Teich, J., Regler, R., Lang, B.: Concurrent topology and routing optimization in automotive network integration. In: Proceedings of Design Automation Conference (DAC), pp. 626–629 (2008)
9. Lim, H.T., Herrscher, D., Waltl, M.J., Chaari, F.: Performance analysis of the IEEE 802.1 Ethernet Audio/Video Bridging Standard. In: Proceedings of the International Conference on Simulation Tools and Techniques (ICST) (2012)
10. Lim, H.T., Krebs, B., Völker, L., Zahrer, P.: Performance evaluation of the inter-domain communication in a switched Ethernet based in-car network. In: Proceedings of the Conference on Local Computer Networks (LCN) (2011)
11. Lim, H.T., Weckemann, K., Herrscher, D.: Performance study of an in-car switched Ethernet network without prioritization. In: Communication Technologies for Vehicles, pp. 165–175. Springer (2011)
12. Steinbach, T., Kenfack, H.D., Korf, F., Schmidt, T.C.: An extension of the OMNeT++ INET framework for simulating real-time ethernet with high accuracy. In: Proceedings of the International Conference on Simulation Tools and Techniques (ICST), pp. 375–382 (2011)
13. Garner, G., Gelter, A., Teener, M.: New simulation and test results for IEEE 802.1as timing performance. In: International Symposium on Precision Clock Synchronization for Measurement, Control and, Communication, pp. 1–7 (2009)

14. Henia, R., Hamann, A., Jersak, M., Racu, R., Richter, K., Ernst, R.: System level performance analysis - the SymTA/S approach. Computers and Digital Techniques, IEE Proc. - 152, 148–166 (2005). doi:10.1049/ip-cdt:20045088.
15. Reimann, F., Kern, A., Haubelt, C., Streichert, T., Teich, J.: Echtzeitanalyse Ethernet-basierter E/E-Architekturen im Automobil. In: GMM-Fachbericht - Automotive meets Electronics (AmE), vol. 64, pp. 9–14 (2010)
16. Rox, J., Ernst, R., Giusto, P.: Using timing analysis for the design of future switched based Ethernet automotive networks. In: Proceedings of Design, Automation and Test in Europe (DATE 12), pp. 57–62 (2012)
17. Imtiaz, J., Jasperneite, J., Han, L.: A performance study of Ethernet Audio Video Bridging (AVB) for industrial real-time communication. In: IEEE Conference on Emerging & Technologies Factory Automation (ETFA), pp. 1–8 (2009)
18. Manderscheid, M., Langer, F.: Network calculus for the validation of automotive Ethernet in-vehicle network configurations. In: 2011 International Conference on Cyber-Enabled Distributed Computing and Knowledge Discovery (CyberC), pp. 206–211 (2011)
19. The Institute of Electrical and Electronics Engineers, Inc.: IEEE Standard for Local and metropolitan area networks-Media Access Control (MAC) Bridges and Virtual Bridged Local Area, Networks (2011)
20. Balarin, F., Chiodo, M., Giusto, P., Hsieh, H., Jurecska, A., Lavagno, L., Passerone, C., Sangiovanni-Vincentelli, A., Sentovich, E., Suzuki, K., Tabbara, B. (eds.): Hardware-Software Co-design of Embedded Systems: The POLIS Approach. Kluwer Academic Publishers, Norwell, MA, USA (1997)
21. Graf, S., Russ, T., Glaß, M., Teich, J.: Considering MOST150 during virtual prototyping of automotive E/E architectures. In: Proceedings of the Automotive meets Electronics (AmE), pp. 116–121 (2012)
22. Kern, A., Schmutzler, C., Streichert, T., Hübner, M., Teich, J.: Network bandwidth optimization of Ethernet-based streaming applications in automotive embedded systems. In: Proceedings of 19th International Conference on Computer Communications and Networks (ICCCN), pp. 1–6 (2010)
23. The Institute of Electrical and Electronics Engineers, Inc.: Local and metropolitan area networks - Part 3: Carrier sense multiple access with collision detection (CSMA/CD) access method and physical layer specifications, IEEE Std 802.3, 2000 edn. (2000)
24. aquintos GmbH: Preevision. http://www.aquintos.com
25. Glaß, M., Lukasiewycz, M., Teich, J., Bordoloi, U., Chakraborty, S.: Designing heterogeneous ECU networks via compact architecture encoding and hybrid timing analysis. In: Proceedings of Design Automation Conference (DAC), pp. 43–46 (2009)
26. Lukasiewycz, M., Glaß, M., Reimann, F., Teich, J.: Opt4J - a modular framework for metaheuristic optimization. In: Proceedings of the Genetic and Evolutionary Computing Conference (GECCO). Dublin, Ireland (2011)

Index

A. Sangiovanni-Vincentelli et al. (eds.), *Embedded Systems Development*,
Embedded Systems 20, DOI: 10.1007/978-1-4614-3879-3,

MIX
Papier aus verantwortungsvollen Quellen
Paper from responsible sources
FSC® C105338

If you have any concerns about our products,
you can contact us on
ProductSafety@springernature.com

In case Publisher is established outside the EU,
the EU authorized representative is:
Springer Nature Customer Service Center GmbH
Europaplatz 3, 69115 Heidelberg, Germany

Printed by Libri Plureos GmbH
in Hamburg, Germany